CLIMATE ACTION ROAD MAP FOR CENTRAL AND WEST ASIA, 2025–2030

DECEMBER 2024

ASIAN DEVELOPMENT BANK

ADB

6 ADB Avenue, Mandaluyong City, 1550 Metro Manila, Philippines
Tel +63 2 8632 4444; Fax +63 2 8636 2444
www.adb.org

ISBN 978-92-9277-068-6 (print); 978-92-9277-069-3 (PDF); 978-92-9277-070-9 (ebook)
Publication Stock No. SGP240577-2
DOI: http://dx.doi.org/10.22617/SGP240577-2

Notes:
In this publication, "$" refers to United States dollars.
ADB recognizes "Kyrgyzstan" as the Kyrgyz Republic.

Cover design by Mike Cortes.

All photos by ADB.

CONTENTS

TABLES, FIGURES, AND BOXES

Tables

Figures

Boxes

MESSAGE FROM THE DIRECTOR GENERAL

It is with great pride that I introduce the *Climate Action Road Map for Central and West Asia, 2025–2030*. This document represents a pivotal step forward to embrace the climate shift under the new operating model of the Asian Development Bank (ADB).

As the effects of climate change continue to escalate, manifesting through devastating floods, heat waves, and prolonged droughts, ramping up climate operations in Central and West Asia region is urgently needed more than ever. Developing member countries (DMCs) in the region face common yet diverse climate challenges—from managing scarce water resources and glacial melting, addressing low agricultural productivity, tackling social vulnerabilities to transitioning away from fossil fuel dependence while maintaining economic stability. Immediate, targeted and comprehensive solutions are crucial to support vulnerable populations, ecosystems, economies that bear the brunt of these climate impacts.

As the climate bank and steadfast partner of Asia and the Pacific, ADB remains deeply committed to addressing these pressing challenges. Guided by Strategy 2030 and the recently launched *Climate Change Action Plan, 2023–2030*, this climate action road map lays out a clear and actionable framework to enhance climate resilience, foster sustainable economic growth, and deliver transformative change across key sectors. This road map underscores the Central and West Asia Department's dedication to ensure that our operations are responsive to the climate shift through the following:

- Promotion of a collaborative, programmatic, and multisectoral approach in DMCs
- Mobilization of up to $25 billion climate finance from sovereign and nonsovereign operations by 2030
- Identification of country, cross sectoral, and regional climate priorities to create synergies
- Prioritization of grant and concessional fund raising of climate finance resources through cofinancing and partnerships
- Mobilization of additional climate actions, including the use of new modalities and innovative products to expand climate operations

The *Climate Action Road Map for Central and West Asia, 2025–2030* champions regional cooperation through different platforms including Central Asia Regional Economic Cooperation (CAREC) Program. It is expected to facilitate shared learning and integrated responses to climate challenges in the region. Furthermore, this road map will be supplemented by climate action plans of countries in Central and West Asia to guide programming processes that can help achieve climate priorities in the country partnership strategies.

The road map is a call for regional action and programmatic approach to deliver transformative and impactful climate outcomes. Together, let us seize this opportunity to chart a path toward a sustainable, resilient, and inclusive future for generations to come in Central and West Asia.

Yevgeniy Zhukov
Director General,
Central and West Asia Department, ADB

ACKNOWLEDGMENTS

The preparation of this *Climate Action Roadmap for Central and West Asia, 2025–2030* was spearheaded by Kathleen Anne Coballes, former Asian Development Bank (ADB) staff Malte Maass, and Jesper Klindt Petersen, with invaluable support from Branko Wehnert and Francis Heil under the technical assistance "Delivering a Climate Change Strategy for Central and West Asia." The team benefitted from the guidance of ADB's Central and West Asia Department (CWRD) senior management, primarily Director General Yevgeniy Zhukov and Deputy Director General Xiaohong Yang.

Many thanks are given to the regional and country management teams and the Climate Change, Resilience, and Environment Cluster under the Climate Change and Sustainable Development Department for the steer and peer review, namely Director General Bruno Carrasco, Toru Kubo, Noelle O'Brien, Alexandra Vogl, Arghya Sinha Roy, Kate Hughes, Esmyra Javier, Nathan Rive, and Leo Kris Palau. The team also extends gratitude to Fei Yu and Declan Magee for their inputs and advice.

The team is also grateful for the inputs of the One ADB team family across departments and sectors led by Senior Directors, Directors, and Country Directors who provided feedback during the consultations and on the versions of the road map and its accompanying documents.

The CWRD country focals from sectors and resident missions provided substantial insights and strong support, and were critical anchors during country missions, namely: Mohammad Hanif Ayubi, Vardan Karapetyan, Yagut Ertenliche, Ekaterine Mikadze, Kenzhekhan Abuov, Samat Aldeev, Ainagul Amanova, Hiz Jamali, Asif Turangzai, Ganjina Fazilova, Serdar Charyyev, Begzod Djalilov, Nizomiddin Rakhmanov, Feruza Insavalieva, Maria Pia Ancora, Jinqiang Chen, Giacomo Giuseppe, Xin Long, Ayesha De Lorenzo, Anjum Israr, Yongkeun Oh, Rika Idei, Cha-Sang Shim, Michael Anyala, Johannes Eberhard Vogel, Kelly Hewitt, Matthias Leitner, Kiyoshi Taniguchi, Linda Faustina Arthur, Mary Claribel Tiu Villarico, Abhishek Hegde, and Ting-Ying Lin.

The team would also like to extend its appreciation for the inputs and support from Roberta Gerpacio, Carizon Espenesin, Melissa Alipalo, Iskandar Abdullaev, Marissa Castillo, Jenny Lyn Hernandez, Michelle Spillar, Zak Baillie, Athena Eftychiou, Panagiota Giannouli, Cara-Marie O'Keeffe, Hannah Patrick, and Caolan Leneghan.

Timely assistance on coordination and administrative matters was provided by Anna Belinda Casiano, Eathel Jean Clar, Cielo Dalde, and Michelle Yrreverre.

The publication of this road map was facilitated with Peter Fredenburg, editor; Samantha Brown, copyeditor; Jonathan Yamongan, Lawrence Casiraya, and Jess Macasaet, in close coordination with Kiyoshi Taniguchi, Eliot John Avena, and the Department of Communications and Knowledge Management. Mike Cortez provided the cover design.

ABBREVIATIONS

ADB	—	Asian Development Bank
AFD	—	Agence Française de Développement
AIIB	—	Asian Infrastructure Investment Bank
CAP	—	climate action plan
CAREC	—	Central Asia Regional Economic Cooperation
CBAM	—	Carbon Border Adjustment Mechanism
CCAP	—	Climate Change Action Plan, 2023–2030
CPS	—	country partnership strategy
CWA	—	Central and West Asia
DMC	—	developing member country
EBRD	—	European Bank for Reconstruction and Development
EIB	—	European Investment Bank
EU	—	European Union
GCF	—	Green Climate Fund
GDP	—	gross domestic product
GHG	—	greenhouse gas
IFI	—	international finance institution
m	—	meter
MDB	—	multilateral development bank
NAP	—	national adaptation plan
ND-GAIN	—	Notre Dame Global Adaptation Initiative
NDC	—	nationally determined contribution
PBL	—	policy-based loan
PPP	—	public–private partnership
WBG	—	World Bank Group

EXECUTIVE SUMMARY

A. Challenges for Climate Resilience and Low-Carbon Development in Central and West Asia

Countries in Central and West Asia (CWA) face significant physical and transition climate risks. Rising temperatures, changing precipitation patterns, and climate extremes pose major challenges for communities and economies across the region, threatening food, water, and energy security. Vulnerable populations—particularly in Afghanistan and Pakistan, mountainous regions of the Caucasus and Central Asia, and high-density urban areas—are highly susceptible to climate hazards. Devastating storms, droughts, heat waves, and landslides are increasing and already having major economic impacts, such as catastrophic floods in Pakistan in 2022 and in Afghanistan in 2024. Transition risks are significant for major fossil fuel producers: Azerbaijan, Kazakhstan, Turkmenistan, and Uzbekistan. Asian Development Bank (ADB) developing member countries (DMCs) must navigate the shift to low-carbon economies while maintaining economic growth and ensuring a just transition. Additionally, emerging policies in Europe and potentially elsewhere pose future risks for several countries that export products with high embodied carbon, particularly Azerbaijan, Kazakhstan, and Uzbekistan. Balancing sustainable economic growth with decarbonization and climate resilience is crucial for the region.

Climate risks pose adverse impacts on key sectors, but there are also opportunities to reduce greenhouse gas (GHG) emissions and deliver green growth. Agriculture faces significant threats from higher temperatures, drought, and other extreme weather, while being a major contributor to GHG emissions. Opportunities exist in regenerative agriculture and methane emission reduction, particularly in Pakistan. Water security is a critical concern, with looming overexploitation and climate-induced variability exacerbating scarcity. Enhancing water efficiency and the sustainable management of resources is essential. The energy sector faces reduced efficiency and increased demand due to heat waves, and drought poses risks to hydropower, alongside challenges stemming from the region's reliance on fossil fuels. Transport infrastructure faces risks from heat stress, landslides, and flooding, while vehicles that burn fossil fuels remain one of the largest sources of GHG emissions. Advances in electric vehicles and increased public transport and nonmotorized travel offer many benefits. Urban areas face heightened risks from heat and flooding, necessitating resilient and green urban planning and reduced carbon emissions from buildings. Health impacts include increased heat-related illness and waterborne disease. Comprehensive disaster risk management is essential to safeguard lives and livelihoods. The private sector has a critical role to play in reducing GHG emissions and enhancing the sustainability of operations, and the finance sector must enable resilient investment and direct finance away from climate-harming activities. Policies and actions taken to address risks and opportunities must deliver a fair transition toward low-carbon economies and resilient societies.

DMCs in CWA have made climate commitments with differing priorities toward mitigation and adaptation. The region contributes some 1.5% of global GHG emissions, and several DMCs face challenges in meeting their nationally determined contribution (NDC) commitments. Economic growth is generally tied to rising GHG emissions, and several regional economies are fossil fuel exporters. Armenia, Georgia, and the Kyrgyz Republic aim to achieve net zero by 2050, and Kazakhstan has set a net-zero target by 2060. National long-term low-carbon climate strategies, crucial for the green economy transition, are being developed, and some countries already have detailed action plans. Adaptation is a key focus to address water, food, and energy security and extreme climate impacts from melting glaciers, drought, heat waves, flooding, and debris flows. Pakistan stands out in terms of the scale of need across both mitigation and adaptation, and Turkmenistan and Uzbekistan follow. An assessment of climate policies and consultative discussions with government representatives suggest mitigation should be the greater priority for Kazakhstan, but adaptation a greater priority for Afghanistan and Tajikistan.

A significant gap exists in meeting the climate finance needs of the region. It is estimated that up to $100 billion climate finance is needed annually in CWA to implement DMC mitigation and adaptation policies. The energy sector has received the most climate finance in recent years, and this needs to expand to other sectors, alongside increasing adaptation finance, particularly for transport, agriculture, urban development, water, and health.

B. ADB Response to Central and West Asia's Climate Needs

As the climate bank for Asia and the Pacific, ADB is committed to supporting DMC efforts to tackle the impacts of climate change. ADB has committed to deliver $100 billion in climate finance from its own resources from 2019 to 2030 for adaptation, resilience, and GHG reduction. Guided by ADB's Strategy 2030 with an operational priority to tackle climate change, build climate and disaster resilience, and enhance environmental sustainability, ADB's *Climate Change Action Plan, 2023–2030* (CCAP) was launched at the 2023 Conference of the Parties of the United Nations Framework Convention on Climate Change (COP28). It guides how ADB operates, delivers, and engages as it responds to the climate needs and priorities of DMCs. It updated ADB's target for climate finance to reach 50% of total committed financing volume by 2030. To operationalize the CCAP at the regional level, ADB has developed the *Climate Action Road Map for Central and West Asia, 2025–2030*, setting out the areas of focus for ADB operations in the region across key themes and sectors.

This road map promotes ADB's shift to provide more client-centered, cross-sector, and integrated solutions to climate and development challenges in CWA. ADB will deliver strategic interventions upstream (e.g., enhancing DMC's climate policy frameworks), midstream (e.g., integrating climate change into DMC sector plans and budgets), and downstream (e.g., delivering climate programs and projects) to help DMCs achieve their climate goals. Strategic interventions are clustered into four thematic areas:

(i) **Governance and public sector reform.** Help countries establish policies and regulatory frameworks for low-carbon energy investments and emission reduction, as well as support governance reform to enhance disaster risk management and water security.

(ii) **Green and resilient economic growth.** Promote diversified, low-carbon, and resilient economies across the region. This includes investment in solar and wind energy, developing cleaner industrial processes, and decarbonizing urban transportation systems.
(iii) **Resilient and inclusive societies.** Enhance social resilience, particularly in urban areas vulnerable to climate risks like heat waves and flooding. ADB will work to better integrate climate resilience into health, education, and social service infrastructure.
(iv) **Sustainable management of natural resources and the environment.** Promote climate-smart agriculture and sustainable water management, especially in water-stressed countries. ADB will emphasize ecosystem conservation and transboundary water cooperation, which is critical to managing shared water resources in the region.

The road map aligns ADB operations with DMC climate priorities. It aims to make ADB operations more responsive to climate needs and opportunities in CWA DMCs. It guides how climate change outcomes can be strategically mainstreamed into programs and projects at the national and regional level by promoting priority climate actions across sectors. The road map will be accompanied by climate action plans (CAPs) to provide country-specific interventions that ensure a targeted and tailored approach to addressing climate change across the CWA region.

C. Accelerating Climate Finance in Central and West Asia

ADB has made significant climate investments across sectors in CWA. From 2019 to 2023, ADB sovereign and nonsovereign climate financing in CWA delivered over $4.9 billion, with 43% allocated to adaptation and 57% to mitigation. ADB nonsovereign private sector finance to DMCs in the region accounted for 10% of total finance, primarily mitigation measures. Pakistan and Uzbekistan together recorded the highest share of ADB climate finance in CWA, at 77%, while ADB had much smaller operations in Azerbaijan, Kazakhstan, and Turkmenistan. Despite the dominance of infrastructure sectors such as transport, energy, and water in ADB's CWA portfolio, diversification into education, health, public sector management, and finance is growing. This must be accelerated as ADB shifts its operations to be more climate responsive and transformative.

ADB aspires to deliver to CWA up to $25 billion in sovereign and nonsovereign climate financing by 2030. ADB will continue to broaden blended finance combining concessional loans, guarantees, and grants to support climate-responsive priorities in CWA DMCs through the One ADB approach. It will ensure that climate financing expands to key vulnerable and emission-intensive sectors across DMCs to meaningfully deliver transformative climate actions. Out of this climate investment, up to $10 billion is for adaptation in key vulnerable sectors.

ADB will ramp up the delivery of additional climate finance and enable climate investment in CWA. Addressing the substantial climate finance gap across CWA requires a multifaceted approach to leverage available resources effectively. ADB has committed to scaling up climate finance and mobilizing private and public investment through a range of innovative tools and partnerships including the following mechanisms:

(i) **Innovative Finance Facility for Climate in Asia and the Pacific.** This is ADB's flagship program designed to leverage guarantees and risk-sharing mechanisms to crowd in private finance. By de-risking projects through sovereign guarantees and partial credit guarantees, ADB can amplify available funding for large-scale, climate-resilient projects. This facility aims to unlock investment in key sectors like renewable energy, infrastructure, and water management, which are much needed across the region.

(ii) **Blended and concessional finance.** ADB is expanding its use of blended finance by working with multilateral partners and global climate funds, including the Green Climate Fund and the Global Environment Facility. These partnerships will leverage ADB resources through concessional loans and grants to support climate adaptation and mitigation efforts, especially in heavily indebted countries such as the Kyrgyz Republic and Pakistan, and in Tajikistan, where climate vulnerability is very high. In addition, ADB will prioritize financing for adaptation projects and make grant resources available to the most vulnerable countries in CWA.

(iii) **Utilizing a range of financing modalities to deliver climate action.** ADB will explore opportunities to deliver policy-based loans, results-based loans, sector development programs, and other financing modalities to deliver climate impact. It will promote "climate policy-based loans" to develop climate policy frameworks in DMCs, support the integration of climate change into fiscal and sector planning, and strengthen the regulatory environment for greater involvement of the private sector in financing climate action.

(iv) **Public–private partnerships.** ADB will actively promote these partnerships to facilitate private sector involvement in financing climate projects. By improving the regulatory environment and offering guarantees to reduce perceived risk, ADB aims to foster investment in renewable energy, low-carbon transport, and sustainable urban and social infrastructure. This approach is particularly relevant where transitioning away from fossil fuel dependency requires substantial private capital.

(v) **Local currency financing.** To mitigate currency risk, ADB will explore local currency financing, particularly in countries with volatile exchange rates. By offering loans in local currencies, ADB can provide more stable financial terms and thus encourage domestic investment in climate action.

D. Enhancing Regional Cooperation and Integration

ADB will enhance regional cooperation on climate action, environmental sustainability, and resilience. It will continue to utilize the Central Asia Regional Economic Cooperation (CAREC) platform, particularly its Working Group on Climate Change and Water Pillar. CAREC provides a suitable platform to take forward regional capacity-building projects and knowledge-sharing initiatives identified in this road map. The *Climate Action Road Map for Central and West Asia, 2025-2030*, country climate action plans (CAPs), the CAREC Climate Change Vision, and the forthcoming CAREC Climate Change Action Plan are well aligned and mutually reinforcing. This road map encompasses both regional and national interventions, leveraging CAREC's established role as a regional cooperation platform.

E. Implementing and Monitoring the Road Map

The road map will guide ADB operations in CWA and provide strategic direction to ADB thematic and sector groups and resident missions when shaping climate interventions in the region. It will support the operationalization of country CAPs and support the delivery of ADB's country partnership strategy in each DMC. Strategic adaptation and mitigation actions for each DMC are detailed in the country CAP. The CAPs will be monitored following the review and updating process of country partnership strategies, which will support effective implementation.

1 INTRODUCTION

1. This *Climate Action Road Map for Central and West Asia, 2025–2030* outlines a comprehensive approach to addressing the region's climate challenges, supporting Asian Development Bank (ADB) developing member countries (DMCs) in the region as they strive for resilient, low-carbon development. This road map aligns with ADB's broader climate ambitions under its Strategy 2030 and *Climate Change Action Plan, 2023–2030*, under which ADB has committed to deliver $100 billion in cumulative climate finance from its own resources from 2019 to 2030, by which time 50% of the total committed financing volume should be climate finance.

Purpose and Scope

2. The primary objective of the road map is to provide a cohesive framework of investment opportunities in Central and West Asia (CWA) that is tailored to the climate policy context and climate finance needs of DMCs in the region. The framework enhances climate resilience, supports green economic growth, and facilitates a transition to low-carbon development across CWA (Figure 1). It offers sector- and thematic-specific guidance and recommendations for scaling up climate finance and fostering regional cooperation through, for instance, the Central Asia Regional Economic Cooperation (CAREC) platform. It does so by encouraging policy reform; investing in climate mitigation and adaptation; and strengthening governance, partnerships, and private sector involvement. The road map is designed to be flexible, allowing ADB operational teams to adapt it to DMC national climate priorities. Country-specific climate action plans (CAPs) are developed to support road map operationalization in individual countries. The road map aims to encourage DMCs, through the CAPs, to align their priorities with the region's broader objectives.

Structure of the Road Map

3. Each part of the road map offers insights into critical aspects of climate risks and resilience, guiding policy decisions and investments through a structured approach.

4. **Positioning.** This section provides a summary of CWA's diverse geography, socioeconomics, and context. Countries in the region face varying degrees of vulnerability to climate change with different impacts on economic and social stability. The geopolitical importance of the region, especially as an energy supplier, adds complexity to climate challenges. This overview helps frame the broader context, setting the stage for a more detailed examination of specific climate risks. It also presents ADB's climate ambitions and targets. ADB's commitment to climate financing goes beyond financial targets, emphasizing partnerships and the bank's catalytic role in mobilizing private sector investment and

Figure 1: Theory of Change of the Climate Action Road Map for Central and West Asia

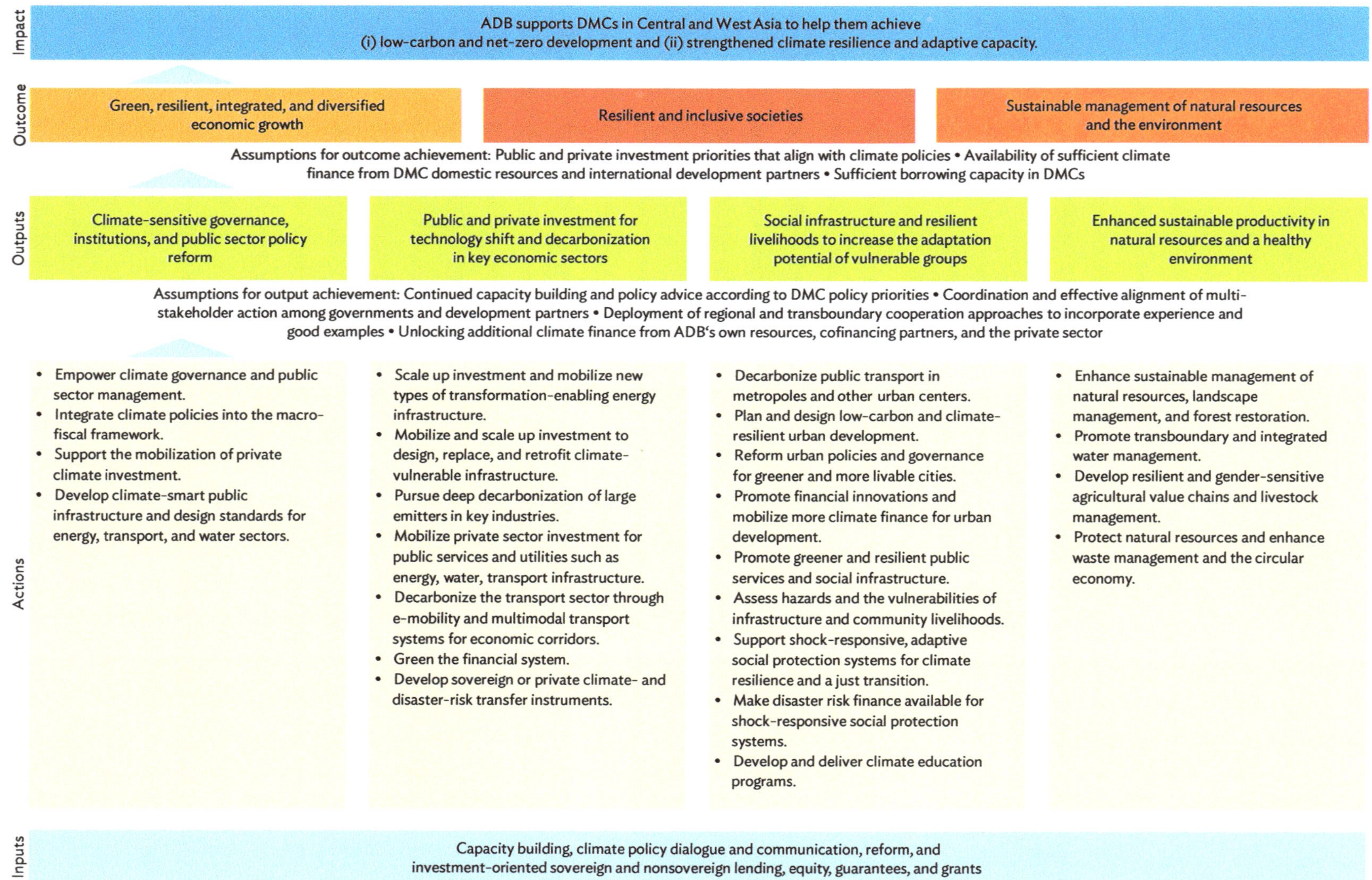

ADB = Asian Development Bank, DMC = developing member country.

Note: This theory of change reflects the structure of the strategic directions of the *Climate Action Road Map for Central and West Asia, 2025–2030* at the outcome level, except the pillar "climate-sensitive governance, institutions, and public sector policy reform," which is subordinate to the other three pillars.

international support. ADB is committed to driving transformative change across its DMCs, ensuring that both adaptation and mitigation efforts are prioritized. CWA's climate risks are outlined, with the importance of regional cooperation underscored, particularly in managing transboundary resources such as water and energy. The climate finance gap is also discussed, emphasizing the need for innovative mechanisms to meet the region's climate goals.

5. **Strategic interventions in the road map.** ADB's approach to supporting climate action in CWA is articulated through upstream, midstream, and downstream interventions. Upstream, the focus is on improving policy frameworks and integrating climate considerations into governance structures. Midstream efforts embed climate objectives into institutional and sector planning. At the downstream level, the emphasis shifts to operationalizing high-quality, climate-sensitive projects. The road map offers strategic directions under four thematic areas: governance and public sector reform, green economic growth, inclusive societies, and sustainable natural resource management. Mobilizing additional climate finance, especially from private sources and international funds, remains a priority.

6. **Implementation, monitoring, and reporting.** Effective execution is supported by robust mechanisms for monitoring and adapting interventions. ADB will work closely with CWA DMCs to ensure that climate actions are implemented efficiently and monitored consistently. Feedback loops (e.g., country management teams) are designed to permit adjustments, ensuring that interventions remain aligned with evolving climate risks and needs. Progress will be tracked through transparent reporting systems, ensuring accountability in the allocation of climate finance and the achievement of key milestones.

7. **Appendixes.** Sector-by-sector analysis highlights the most pressing climate risks and opportunities. Agriculture, energy, transportation, water, urban development, health, finance, and the private sector are all assessed for their exposure to climate change. In addition to identifying risks, appendixes explore opportunities for green development, such as renewable energy and sustainable infrastructure investment. They include a detailed presentation of ADB's climate finance by country, based on commitments from 2019 to 2023, as well as available financing modalities.

2 POSITIONING

8. This chapter provides a regional overview for CWA DMCs. It outlines the climate risks the CWA region faces and the national and regional priorities for delivering climate resilience and low-carbon development. The chapter presents the current level of climate finance in CWA from multilateral development banks, other international finance institutions, and private sources, and estimates the climate finance gap for CWA to 2030. It evaluates ADB's climate finance commitments in CWA in the last 5 years and its plans to play a major role in filling the climate finance gap in the next 5 years.

Regional Overview

9. CWA is a diverse region and a geopolitical hotspot. CWA encompasses a vast territory consisting of 10 DMCs with 384 million people.[1] Located in a geopolitical hotspot between major economic and political powers in the People's Republic of China, Europe, India, the Middle East, and Russia, the region has diverse cultures, languages, and geographies. Countries across the region share several similarities in terms of climate vulnerability and hazards, yet each presents significant differences in terms of geography, economy, and population that can both exacerbate climate risk and strain capacity to adapt and smoothly transition to a low-carbon development path.

10. Locations across the region have contrasting geographies and climates. CWA is largely landlocked, except for Pakistan's southern coastline and Georgia's position on the Black Sea. The region includes glaciers, mountains, deserts, forests, seas, large lakes, and major rivers.

- Central Asia is the largest subregion, with a distinctive arid climate in the south and a more temperate climate in the north. The terrain across the vast landscape is largely flat and arid except in the southeast, where the Kyrgyz Republic and Tajikistan are largely mountainous and green, with elevations exceeding 3,000 meters (m) above sea level.
- The South Caucasus subregion, composed of Azerbaijan, Armenia, and Georgia, is notable for its steep landscape and high elevation, with an average elevation exceeding 1,500 m, Armenia being the most elevated. The region's climate is subtropical and humid, with cooler alpine weather in the mountains to the north. This creates large, dense forests and plant cover in the mountains, with more fluvial flood plains and fields for farming in the lower regions of Azerbaijan facing the Caspian Sea.

[1] ADB placed its regular assistance to Afghanistan on hold effective 15 August 2021. This document covers Armenia, Azerbaijan, Georgia, Kazakhstan, the Kyrgyz Republic, Pakistan, Tajikistan, Turkmenistan, and Uzbekistan. Pakistan has the largest population at 252.4 million as of 2024. Central Asia across its five countries has a total population of 78.2 million, while the three countries in the South Caucasus subregion have a population of 17.0 million. Afghanistan recorded a population of 42.6 million as of mid-2024. Source: Worldometer (accessed 20 July 2024).

- Pakistan features sharply contrasting climate and geography. The west and south are arid landscapes with fold mountains, while the east and north are cooler and more temperate, giving way to green pastures fed by the Indus River as it flows from the north toward the Arabian Sea.
- Afghanistan has an arid to semi-arid continental climate with hot summers and cold winters. The country's landscape is predominantly mountainous, with an average elevation of 1,884 m. Afghanistan's main rivers are the Amu Darya in the north, the Helmand in the southwest, and the Kabul River in the east. These rivers play a crucial role in the country's water resources and agricultural irrigation systems.

11. The region's economies are diverse. In 2023, the region had a combined gross domestic product (GDP) exceeding $900 billion.[2] Pakistan and Kazakhstan have the largest economies in the region, though Pakistan's GDP per capita is among the lowest. Central Asia has the largest agriculture sector as a share of its economy, and had a GDP of $446 billion in 2023. Pakistan and South Caucasus both have large service sectors, while Pakistan boasts the largest industrial sector share at 25%. The South Caucasus countries have a combined GDP of $127 billion, over half of which is produced by Azerbaijan. Table 1 provides details about population, GDP, and the Human Development Index across DMCs. Pakistan and Afghanistan score very low on the Human Development Index, having large numbers of poor and vulnerable people, and Tajikistan, the Kyrgyz Republic, Uzbekistan, and Turkmenistan follow.

Table 1: Population, Gross Domestic Product, and Human Development Index in Central and West Asia ADB Developing Member Countries

Country	Population (million)		GDP per Capita 2022 (current $)[c]	Human Development Index[d]
	2022[a]	2050[b]		
Afghanistan	41.1 (26)	76.9–82.0	372	0.462 (193/204)
Armenia	2.8 (64)	2.4–2.8	7,018	0.786 (76/193)
Azerbaijan	10.1 (57)	10.0–11.8	7,762	0.760 (89/193)
Georgia	3.7 (60)	3.1–3.7	6,675	0.814 (60/193)
Kazakhstan	19.6 (58)	23.5–27.5	11,492	0.802 (67/193)
Kyrgyz Republic	7.0 (37)	8.5–10.2	1,655	0.701 (117/193)
Pakistan	235.8 (38)	335.3–397.6	1,589	0.540 (164/193)
Tajikistan	10.0 (28)	13.8–16.5	1,054	0.679 (126/193)
Turkmenistan	6.4 (53)	7.5–8.9	8,793	0.744 (94/193)
Uzbekistan	35.6 (50)	41.6–49.3	2,255	0.727 (106/193)

GDP = gross domestic product.

Note: Relative scale (inverse scale for GDP per capita and Human Development Index):

Very high High Medium Low

[a] Number in parentheses shows the percentage of urban population. Data from World Population Prospects—2022 Revision, 2024 (accessed 20 July 2024).

[b] Low–high projection data from World Population Prospects—2022 Revision, 2024 (accessed 1 July 2024).

[c] Data from World Bank Group. GDP (current US$) (accessed 6 October 2024).

[d] In parentheses, global rank over the number of countries considered. Data from United Nations Development Programme. 2024. Human Development Report 2023/2024: Breaking the Gridlock: Reimagining Cooperation in a Polarized World.

[2] World Bank Group. GDP (current US$—Europe & Central Asia (accessed 20 August 2024).

Country Climate Commitments in Central and West Asia

12. DMCs across CWA have committed to action on climate change. They have set various targets under their updated nationally determined contributions (NDCs), national adaptation plans (NAPs), and long-term low-emission development strategies. Target years run from 2030 to 2060 (Table 2).

Climate Risks and Impacts in Central and West Asia

Physical Climate Risk Overview

13. CWA faces extreme heat, melting glaciers, increased debris flows, drought, desertification, flooding, cyclones, and sea level rise because of climate change. Average temperatures are projected to increase by up to 7°C toward the end of the 21st century, compared to 1981–2010.[3] The highest temperature increases are projected in the mountains due to elevation-dependent warming. In the plains, the number of hot days above 35°C is projected to increase by up to 80 days per year, posing significant heat risk to people and infrastructure. Maximum 1-day precipitation is projected to increase in some areas by up to 30%, increasing risks from flooding, landslides, and mudflows. A staggering 50%–70% increase is projected in 1-day precipitation in monsoon-exposed parts of Pakistan. While some areas show a projected increase in annual precipitation by up to 20%, drought risk will also increase in some areas, with up to 30 more consecutive dry days per year. Alongside this, desertification is increasing along with sandstorms in some areas, compounding air quality issues. Pakistan's coastline is projected to experience a rise in sea level, exacerbating coastal erosion, saltwater intrusion, and flooding, particularly in the Indus Delta.[4] Table 3 provides an overview of the severity of climate hazards across CWA DMCs[5] and their adaptation readiness as measured under the Notre Dame Global Adaptation Initiative (ND-GAIN).[6]

3 Climate Model Intercomparison Project 6 Ensemble. Climate scenarios: SSP1-RCP2.6, RCP4.5, and SSP5-RCP8.5. See Climate Data Store.

4 J. H. Weeks. 2023. *Sea-Level Rise in Pakistan: Recommendations for Strengthening Evidence-Based Coastal Decision-Making.* Hydrology. 10 (11). p. 205.

5 Hazard ratings do not include assessment of exposure or vulnerability, as ratings are not climate risk assessments.

6 ND-GAIN. Country Index (accessed 20 July 2024).

Table 2: Development of Climate Policies in Central and West Asia ADB Developing Member Countries

Country	Updated NDC Target by 2030	Net-Zero Target	NDC Implementation/ Investment Plan	National Climate Strategy Green Transition of the Economy Strategy	Long-Term Strategy	National or Sector Adaptation Plan(s)
Afghanistan	2015, 13.6% reduction in emission levels by 2030 compared to BAU scenario	None	None	Yes, Climate Change Strategy and Action Plan, 2015	Not yet developed	Under development
Armenia	2021, 40% reduction from 1990 emission levels	Yes, 2023; net zero by 2050	Under development	Pending approval	Yes, 2023, 2050 mitigation goal: 2.07 tCO_2e per capita per annum	Yes, 2021. NAP 2021–2025, and several sector adaptation plans[a]
Azerbaijan	2023, 40% reduction from 1990 emission levels by 2050	None	Not yet developed	Not yet developed[b]	Not yet developed	Under development
Georgia	2021, unconditional; 46% reduction from 1990 emission levels[c]	Yes, 2023; net zero by 2050	Action Plan 2024–2025 available[d]	Yes, 2021 and action plan	Yes, 2023	Yes, 2017 Agricultural SAP
Kazakhstan	2023, unconditional; 15% reduction from 1990 emission levels	Yes, 2023; net zero by 2060	Not yet developed	Yes, 2013	Yes, 2020	Not yet developed
Kyrgyz Republic	2021, unconditional; 15.97% reduction below BAU	2050, verbal commitment 2022[e]	Draft available and under public discussion since December 2023	Yes, Phase 1: 2019–2023; Phase 2: 2024–2028 under development	Under development, supported by UNDP	SAPs under preparation, supported by UNDP, GIZ, and ADB
Pakistan	2021, unconditional; 15% reduction from 1990 emission levels	None	Under development and provincial plans	Updated National Climate Change Policy, 2021	Under development	Yes, 2023
Tajikistan	2021, unconditional; 30%–40% reduction from 1990 emission levels	No explicit commitment on a net zero target[f]	Available as draft, not published	Yes, 2023	Not yet developed	National Adaptation Strategy, 2019
Turkmenistan	2022, unconditional; 20% reduction below BAU in 2010	None	Not yet developed	Yes, 2019	Not yet developed	Under preparation
Uzbekistan	2021; 35% reduction from 2010 emission level	None	Under development	National Climate Strategy under preparation; Green Transition of the Economy Strategy: yes, 2022	Under development[g]	Under development

ADB = Asian Development Bank, BAU = business as usual, GIZ = Deutsche Gesellschaft für Internationale Zusammenarbeit, NAP = national adaptation plan, NDC = nationally determined contribution, SAP = sector adaptation plan, tCO_2e = ton of carbon dioxide equivalent, UNDP = United Nations Development Programme.

[a] ADB supported the development of SAP Agriculture and will support SAP Transport. Other existing SAPs for energy, tourism, water, and health.

[b] United Nations Azerbaijan. 2022. *Green Transformation in Azerbaijan.*

[c] European Union Energy Committee. Decision No 2022/02/MC-EnC of the Ministerial Council of the Energy Community.

[d] No publication date. ADB supported the development of SAP Agriculture and will support SAP Transport. Other existing SAPs are for energy, tourism, water, and health.

[e] UNDP. 2022. Kyrgyzstan Continues to Contemplate Long-Term Climate Change Effects. News release. 12 April.

[f] Mitigation pathways for Tajikistan to achieve carbon neutrality by 2050 are described in UNDP and UKAID. 2021. *Mitigation Pathways for Tajikistan to Achieve Carbon Neutrality by 2050.*

[g] Supported by ADB, Agence Française de Développement, and the World Bank Group.

Sources: United Nations Climate Change. Nationally Determined Contributions Registry (accessed 20 July 2024); Climate Watch. Explore Nationally Determined Contributions (accessed 20 July 2024); and ADB.

Table 3: Climate Hazards in Central and West Asia

Country	Drought	Water Stress	Heat	Landslide	Riverine Flood	Coastal Flood	Hailstorm	ND-GAIN Adaptation Score
Afghanistan	High in western, southern, and eastern provinces	High in urban centers	Medium in northern, northeastern, and southern regions	High in north, northeast, and south; medium in central and east regions	High in the north and east	n.a.	Low	Extremely high score = 33.2, 181st out of 185 countries
Armenia	North: medium; south: medium–high	North: low; central: extremely high	Low, except Aras Valley	High in mountain ranges	Medium	n.a.	Medium	Medium score = 56.9, 49th out of 185 countries
Azerbaijan	Medium–high	Extremely high in Kura lowland; low–medium in mountain ranges	Moderate– high	High in mountain ranges	High	n.a.	High	Medium score = 53.3, 65th out of 185 countries
Georgia	Medium	Low–medium, except in Kakheti region, medium–high	Low in mountains; severe on Kolkheti Plain	High in mountain ranges	High	Low on Black Sea coast	Low, except in Tian Shan range	Medium score = 58.5, 38th out of 185 countries
Kazakhstan	Medium–high, highest in north	Extremely high toward the west; low in north and east	Moderate–high in southwest; slight–low in northeast	High in Tien Shan range	Medium	n.a.	Medium	Low score = 59.8, 36th out of 185 countries
Kyrgyz Republic	Medium in mountains; high toward Fergana Valley	Extremely high in north	Low, except near Fergana Valley and Bishkek area	High in all mountain ranges	High	n.a.	High	Medium score = 53.3, 65th out of 185 countries

continued on next page

Table 3 continued

Country	Drought	Water Stress	Heat	Landslide	Riverine Flood	Coastal Flood	Hailstorm	ND-GAIN Adaptation Score
Pakistan	High	High–extremely high except close to Indus River	Severe–very high on plains; moderate–high in mountain ranges	High in Karakoram, Hindu Kush, and Sulaiman ranges	High	High in Sindh	Medium	Extremely high score = 39.6, 150th out of 185 countries
Tajikistan	Low–medium except for high risk in Sughd	High in Sughd, Gorno-Badakhshan, and Khatlon	Low except for moderate–severe in lower areas in Sughd and Khatlon	High in all mountain ranges	High	n.a.	Low	Moderately high score = 47.6, 98th out of 185 countries
Turkmenistan	Medium–high	Extremely high in south; medium in north	Severe in Lebap and Balkan; moderate–high in other parts	High in Kopet Dag range	Medium	n.a.	Low	score = 44.2, 117th out of 185 countries
Uzbekistan	High in south and east; medium in north and west	High–extremely high in most of the country	Severe–very high in south and Syrdaryo; moderate–high in most other parts	High in Tien Shan and Chatkal ranges	Medium	n.a.	High	Medium score = 52.2, 72nd out of 185 countries

Hazard rating:
Severe / Extremely high; High; Medium; Low

n.a. = not applicable, ND-GAIN = Notre Dame Global Adaptation Initiative.

Source: Notre Dame Global Adaptation Initiative. Country Index (accessed 1 July 2024).

14. Vulnerable people and communities face the greatest risk from more frequent and severe impacts. Women, older people, infants, people in poor health or with disabilities, people from lower socioeconomic backgrounds, and members of minority groups are highly vulnerable to climate hazards. In general, communities living in mountainous areas are highly exposed to debris flows, making them particularly vulnerable, and they lack the financial means to adapt and protect themselves from changing hazards. Stormwater flooding, heat waves, and water scarcity are key risks in many urban areas, and widespread river flooding has already caused devastating impacts across the region, such as in Pakistan in 2022. Country-specific disaster risk profiles can help compare economic damage from climate hazards (Table 4). Kazakhstan, the Kyrgyz Republic, and Tajikistan face the greatest costs from flood damage as a percentage of GDP, and potential damage is high in many DMCs.

Table 4: Economic Value Affected by 1:100 Year Flood at Present and in 2080

Country	Affected by 1:100 Year Flood at Present ($ billion [% of 2015 GDP])	Affected by 1:100 Year Flood in 2080 ($ billion [% of 2015 GDP])
Afghanistan	0.5 (3%)	1.5–2.5 (8%–13%)
Armenia	0.7 (6%)	2.0–3.0 (19%–28%)
Azerbaijan	1.0 (2%)	2.0–3.0 (4%–5%)
Georgia	1.0 (10%)	6.0–8.0 (44%–58%)
Kazakhstan	10.0 (11%)	60.0–100.0 (50%–83%)
Kyrgyz Republic	0.4 (7%)	4.0–7.0 (73%–127%)
Pakistan	14.6 (5%)	Data not available[a]
Tajikistan	0.5 (8%)	6.0–9.0 (87%–130%)
Turkmenistan	3.0 (7%)	10.0–30.0 (20%–60%)
Uzbekistan	4.0 (5%)	20.0–30.0 (30%–45%)

Relative scale:

Very high High Medium Low

GDP = gross domestic product.

[a] Assumed to be significantly higher than $14.6 billion (5% of 2015 GDP) at present.

Sources: Data derived from Global Facility for Disaster Reduction and Recovery (GFDRR)/World Bank Disaster Risk country profiles; except for Pakistan, derived from Central Asia Regional Economic Cooperation (CAREC) and ADB. 2021. *The Central Asia Regional Economic Cooperation (CAREC) Program Development Effectiveness Review 2020.*

Transitional Climate Risk Overview

15. Greenhouse gas (GHG) emissions generally continue to rise with economic growth, and there are inequalities in GHG emissions across countries and sectors. In 2022, CWA contributed approximately 1.5% of global GHG emissions. Fossil fuel-producing countries like Azerbaijan, Kazakhstan, Pakistan, Turkmenistan, and Uzbekistan make up 93% of all GHG emissions in the region and are home to 83% of the region's population. Figure 2 shows GHG emissions from CWA DMCs and highlights which sectors have particularly high GHG emissions. Inequalities in GHG emissions are indicated for Kazakhstan and Turkmenistan, with GHG emissions per capita 1.5 times above global average, and for the Kyrgyz Republic, Pakistan, Tajikistan, Turkmenistan, and Uzbekistan, with GHG emissions per dollar of GDP at approximately twice the global average.

Figure 2: Greenhouse Gas Emissions per Country and Sector, Central and West Asia, 2021

MtCO2e in 2021 (per country)	Afghanistan	Armenia	Azerbaijan	Georgia	Kazakhstan	Kyrgyz Republic	Pakistan	Tajikistan	Turkmenistan	Uzbekistan	West Asia	Central Asia	Central and West Asia
Energy	14.9	7.9	42.6	10.9	280.3	9.6	229.1	7.7	173.6	136.6	61.4	607.8	913.2
Electricity/heat	0.2	1.6	14.6	1.0	131.5	2.8	59.0	1.8	21.0	50.9	17.2	208.1	284.5
Building	2.4	1.7	8.4	3.1	39.6	4.3	34.1	1.1	13.7	34.1	13.2	92.8	142.5
Manufacturing/construction	4.6	0.5	2.1	1.7	29.6	0.5	64.5	1.4	0.7	13.0	4.4	45.2	118.6
Transportation	5.4	2.5	7.9	4.3	23.4	1.7	57.5	1.9	12.1	16.9	14.6	56.0	133.6
Other fuel combustion	0.1	0.6	1.3	0.0	2.3	0.0	1.1	1.3	12.1	2.1	2.0	17.9	21.1
Fugitive emissions	2.3	0.9	8.4	0.8	53.8	0.2	12.8	0.1	114.0	19.6	10.1	187.7	212.9
Industrial processes	1.4	0.7	1.6	2.1	5.6	1.3	33.7	2.4	1.5	9.6	4.4	20.5	60.0
Agriculture	14.9	1.7	7.9	1.9	29.0	5.7	208.6	6.9	12.0	38.8	11.5	92.4	327.4
Waste	3.8	0.7	4.4	3.2	5.8	0.8	9.7	0.9	1.5	7.4	8.3	16.4	38.2
Bunker fuels	0.04	0.2	1.4	0.3	1.5	0.0	1.4	0.1	0.1	0.3	1.9	2.0	5.3
Land use change and forestry	0.15	0.0	1.9	0.0	3.0	-3.3	7.0	-0.1	0.0	(3.4)	1.9	(9.7)	(4.5)
Total GHG emissions	35.2	11.2	56.0	18.5	319.3	14.2	489.4	17.9	188.6	189.3	85.7	729.3	1339.6
Relative GHG emissions	3%	1%	4%	1%	24%	1%	37%	1%	14%	14%	6%	54%	
GHG emissions per capita	0.87	3.97	5.3	4.9	16.72	2.09	2.11	1.83	29.73	5.41			
GHG emissions per GDP	2.47	798.34	996.1	974.97	161.2	1530.07	1400.27	1993.69	3770.06	2715.91			
Population 2024 (million)	42.6	2.8	10.5	3.7	19.8	6.8	245.2	10.3	6.6	35.7	17.0	121.8	384.0
Population 2024 regional share (%)	11	1	3	1	5	2	64	3	2	9	4	32	
Population growth (%)	2.68	-0.71	0.32	-0.19	1.34	2.35	1.9	2.23	2.05	1.96			

GDP = gross domestic product, GHG = greenhouse gas, MtCO2e = metric ton of carbon dioxide equivalent.

Note: The Climate Watch dataset includes all sectors and GHGs. To ensure comparability of data across countries, it deliberately does not use countries' official inventories reported to the United Nations Framework Convention on Climate Change.

Source: Climate Watch. Historical GHG Emissions (accessed 23 October 2024).

16. GHG reduction commitments will be challenging to achieve for many countries, and accelerated decarbonization is urgent. Most of the region's DMCs are post-Soviet economies that experienced a dramatic drop in GHG emissions in the early 1990s after the collapse of the Soviet Union. Since then, their annual GHG emissions have increased in line with economic growth. An analysis of the risk of DMCs missing their NDC targets has been undertaken, based on estimating their GHG emissions in 2030 and projecting forward the growth or reduction rate of their annual GHG emissions from 2017 to 2021 (Figure 3). According to the GHG projections, Turkmenistan, Kazakhstan, Uzbekistan, and

Figure 3: Difference Between Nationally Determined Contribution 2030 Greenhouse Gas Targets and Projected 2030 Greenhouse Gas Emissions Reductions in Central and West Asia, Based on Rate of Decarbonization, 2017–2021 (%)

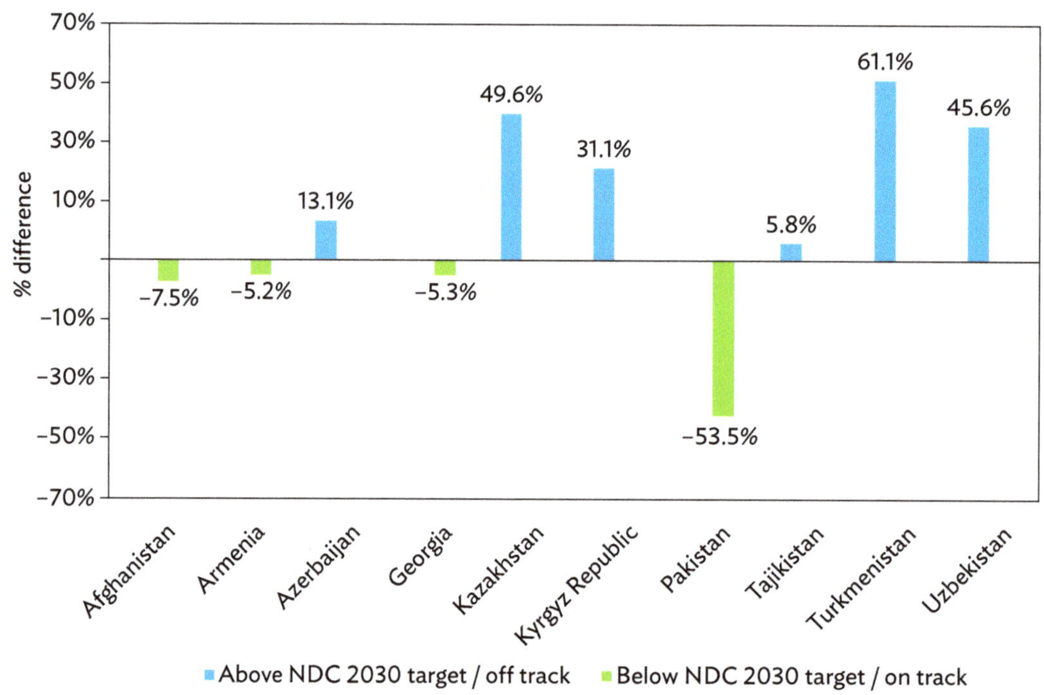

DMC = developing member country, GHG = greenhouse gas, NDC = nationally determined contribution.

Notes:

1. GHG emissions data from Climate Watch Data (accessed 23 October 2024).

2. As each DMC has a different NDC target and different baseline for measuring the target, comparison across countries is not a direct comparison in GHG reduction, but rather a comparison of being on track or off track in meeting NDC commitments. The comparison does not show where some DMCs have set more ambitious targets than others.

3. The analysis has limitations given the impact of coronavirus disease (COVID-19) on GHG emissions, and it is possible DMCs could accelerate their decarbonization beyond 2022 to ensure they meet their NDC targets. Analysis nonetheless provides a useful indication of which DMCs were on or off track as of 2022.

4. Azerbaijan's 2030 intended NDC target—enunciated in 2017 as a 35% reduction in GHG emissions by 2030 based on 1990 emissions—has been used for this analysis to allow for comparison with other countries up to 2030. This 2030 target was removed in Azerbaijan's 2023 NDC; the target is now a 40% reduction in GHG emissions by 2050 based on 1990 emissions.

5. Unlike the other CWA countries whose NDC targets are based on absolute GHG emissions expressed as metric tons of carbon dioxide equivalent, Uzbekistan's NDC target is based on emissions intensity per dollar of gross domestic product. For Uzbekistan, the projection of GHG in 2030 is based on projecting forward the change in emissions intensity from 2017 to 2021, and therefore assumes the same rate of gross domestic product growth to 2030.

Source: Asian Development Bank.

the Kyrgyz Republic are more than 30% off track from their 2030 NDC targets. Azerbaijan and Tajikistan are less than 15% off track. Countries off track face the challenge of maintaining economic growth while rapidly reducing and, if possible, reversing the growth of GHG emissions through decarbonization measures and intensified investment in carbon sinks. Afghanistan, Armenia, and Georgia are projected to meet their NDC targets. Pakistan is projected to have a very low risk of missing its 2030 NDC target, with its GHG emissions projected to be more than 50% below the 2030 target. Countries should focus on reducing their emissions intensity so that they remain on track to meet their 2030 commitments while allowing economic growth with relatively modest increases in GHG emissions.

17. Several DMCs face significant transition risks due to high production and/or consumption of fossil fuels, as well as exports of high-carbon products. As the world transitions away from fossil fuels and high-carbon products, and as major importers bring in policies to account for the carbon in products, countries could face potential reductions in export income and higher costs for energy. One such policy is the European Union (EU) Carbon Border Adjustment Mechanism (CBAM), which is currently being implemented and will come into full force in 2026.[7] The United Kingdom is also considering bringing in a CBAM in 2027. Initially, the EU CBAM will apply to imports of high-carbon cement, iron and steel, aluminum, fertilizers, electricity, and hydrogen. Table 5 sets out a high-level analysis of CWA DMCs toward relevant transitional risks from fossil fuel production and consumption, as well as their exposure to the EU CBAM policy based on the DMC's level of exports to the EU and the carbon intensity of the DMC's energy supply and industry. High production of fossil fuels implies DMCs are more exposed to transition risks, due to the dependency of their economies on consuming and exporting fossil fuels. Similarly, DMCs with high exports of carbon-intensive products are also more exposed to transition risks affecting their exports and their economies.

Table 5: Relative Assessment of Transition Risks Related to Fossil Fuel Production and Consumption and Export of High-Carbon Products

Country	High-Level Assessment of Transition Risks Related to Fossil Fuel		High-Level Assessment of Transition Risks Related to Exposure to EU CBAM, and Key Exported High-Carbon Products
	Production 2022 (quadrillion Btu)[a]	Consumption 2022 (quadrillion Btu)	
Afghanistan	**Low** Coal = 0.1 Natural gas = <0.1 Petroleum = <0.1	**Low** Coal = <0.1 Natural gas = <0.1 Petroleum = <0.1	**Low** (moderate–high dependency on fossil fuels with low EU exports) Key exports: mineral and chemical products. The EU is Afghanistan's fifth-biggest export market (3.6% in 2023).[b]
Armenia	**Low** Coal = <0.1 Natural gas = 0 Petroleum = 0	**Low** Coal = <0.1 Natural gas = <0.1 Petroleum = <0.1	**Moderate** (moderate dependency on fossil fuels with high–moderate EU exports) Key exports: metals (iron, steel), minerals. The EU is Armenia's third-biggest export market (8.4% in 2023).[c]

continued on next page

[7] European Commission. Carbon Border Adjustment Mechanism.

Table 5 continued

Country	High-Level Assessment of Transition Risks Related to Fossil Fuel		High-Level Assessment of Transition Risks Related to Exposure to EU CBAM, and Key Exported High-Carbon Products
	Production 2022 (quadrillion Btu)[a]	Consumption 2022 (quadrillion Btu)	
Azerbaijan	**Moderate** Coal = 0 Natural gas = 1.3 Petroleum = 1.4	**Moderate** Coal = <0.1 Natural gas = 0.4 Petroleum = <0.1	**High** (high dependency on fossil fuels with high EU exports) Key exports: oil, gas fuel, chemicals, petrochemicals. The EU is Azerbaijan's biggest export market (64.4% in 2023).[d]
Georgia	**Low** Coal = <0.1 Natural gas = <0.1 Petroleum = <0.1	**Low** Coal = <0.1 Natural gas = 0.1 Petroleum = <0.1	**Low** (moderate–low dependency on fossil fuels with moderate EU exports) Key exports: electricity, minerals, chemicals. The EU is Georgia's third-biggest export market (11.6% in 2023).[e]
Kazakhstan	**High** Coal = 2.7 Natural gas = 1.1 Petroleum = 3.9	**High** Coal = 2.0 Natural gas = 0.8 Petroleum = 0.6	**High** (high dependency on fossil fuel with high EU exports) Key exports: fuel oil and gas, minerals, chemicals, metals (iron, steel). The EU is Kazakhstan's biggest export market (37.1% in 2023).[f]
Kyrgyz Republic	**Low** Coal = <0.1 Natural gas = <0.1 Petroleum = <0.1	**Low** Coal = <0.1 Natural gas = <0.1 Petroleum =<0.1	**Low** (moderate–low dependency on fossil fuels with low EU exports) Key exports: fuels, metals, minerals. The EU is Kyrgyz Republic's ninth-biggest export market (1.0% in 2023).[g]
Pakistan	**Moderate** Coal = 0.3 Natural gas = 1.1 Petroleum = 0.2	**Moderate** Coal = 0.8 Natural gas = 1.4 Petroleum = 0.9	**Low** (moderate dependency on fossil fuels with low EU exports) Key exports: metals (copper, zinc, lead). The EU is Pakistan's biggest export market (32.1% in 2023). However, most of these exports, such as textiles, are not currently included in the EU CBAM policy.[h]
Tajikistan	**Low** Coal = <0.1 Natural gas = <0.1 Petroleum = <0.1	**Low** Coal = <0.1 Natural gas = <0.1 Petroleum = <0.1	**Moderate** (moderate–high dependency on fossil fuels with moderate EU exports) Key exports: gas fuels, petrochemicals, chemicals. The EU is Tajikistan's third-biggest export market (10.9% in 2023).[i]

continued on next page

Table 5 continued

Country	High-Level Assessment of Transition Risks Related to Fossil Fuel		High-Level Assessment of Transition Risks Related to Exposure to EU CBAM, and Key Exported High-Carbon Products
	Production 2022 (quadrillion Btu)[a]	Consumption 2022 (quadrillion Btu)	
Turkmenistan	**High** Coal = 0 Natural gas = 3.2 Petroleum = 0.5	**Moderate** Coal = 0 Natural gas = 1.5 Petroleum = 0.3	**Moderate** (moderate–high dependency on fossil fuels with high–moderate EU exports) Key exports: industrial goods, metals, chemicals. The EU is Turkmenistan's third-biggest export market (8.9% in 2023).[j]
Uzbekistan	**Moderate** Coal = <0.1 Natural gas = 1.7 Petroleum = <0.1	**Moderate** Coal = 0.1 Natural gas = 1.6 Petroleum = 0.2	**High** (high dependency on fossil fuels with high EU exports) Key exports: cement, metals, steel, chemicals. The EU is Uzbekistan's fifth-biggest export market (4.9%) in 2023, and many of these exports would be subject to the EU CBAM policy.[k]

Btu = British thermal unit, CBAM = Carbon Border Adjustment Mechanism, EU = European Union.

[a] Data sourced from United States Energy Information Administration.

[b] European Commission, Directorate-General for Trade. 2024. *European Union. Trade in Goods with Afghanistan.*

[c] European Commission, Directorate-General for Trade. 2024. *European Union, Trade in Goods with Armenia*.

[d] European Commission, Directorate-General for Trade. 2024. *European Union, Trade in Goods with Azerbaijan.*

[e] European Commission, Directorate-General for Trade. 2024. *European Union, Trade in Goods with Georgia.*

[f] European Commission, Directorate-General for Trade. 2024. *European Union, Trade in Goods with Kazakhstan*.

[g] European Commission, Directorate-General for Trade. 2024. *European Union, Trade in Goods with Kyrgyz Republic*.

[h] European Commission, Directorate-General for Trade. 2024. *European Union, Trade in Goods with Pakistan*.

[i] European Commission, Directorate-General for Trade. 2024. *European Union, Trade in Goods with Tajikistan*.

[j] European Commission, Directorate-General for Trade. 2024. *European Union, Trade in Goods with Turkmenistan*.

[k] European Commission, Directorate-General for Trade. 2024. *European Union, Trade in Goods with Uzbekistan*.

Source: Compiled by the Asian Development Bank.

18. Transition risks exist due to potential shifts in policies, markets, technologies, and societal behavior.[8] Labor markets and educational programs need to shift toward new skills and technologies, posing risks to countries with high employment levels in the fossil fuel industry. Government budgets need to adapt to reduced revenue from exports of fossil fuels and high-carbon products, posing risks to public service provision. Social protection systems need to be adjusted for scalability and shock-responsiveness to support those affected by climate disasters. This transformation contributes to the reduction of future physical climate risks and significantly lowers costs overall, but it brings costs forward to the present. Some countries may face challenges attracting sufficient investment to facilitate the transition, due to perceived higher risks and lower returns. Energy prices are likely to increase in the short to medium term, so targeted subsidies or social safety nets are required to ensure vulnerable communities are safeguarded from increasing poverty and social discontent. Technology

[8] ADB. 2023. *Asia in the Global Transition to Net Zero*.

change poses risks, such as reliance on rare earth metals for batteries, and there is uncertainty about the performance of some technologies, notably carbon capture, usage, and storage, and hydrogen energy. Abrupt policy shifts or uncoordinated actions among nations could undermine business confidence and investment. Countries across CWA need to diligently assess the impacts of transition processes and develop conducive regulatory frameworks, ensuring a fair and just transition toward low-carbon, resilient economies and societies.

Overview of Impacts Across Sectors

19. Climate change and the transition to low-carbon, resilient economies pose challenges and opportunities across sectors. Agriculture faces risks of reduced harvests and worsened food insecurity, while livestock remains a major GHG emitter and will be exposed to health risks under climate change. Opportunities exist for regenerative agriculture and utilizing methane. Energy security is called into question by a transition away from fossil fuels, and this challenge must be balanced with a huge surge in demand for electricity and the need to manage physical climate impacts on generation and distribution infrastructure. Opportunities in renewable energy—particularly hydropower in Georgia, the Kyrgyz Republic, and Tajikistan—have significant potential to provide electricity to regional power grids. Innovations like hydrogen power and new heating technologies will open new economic opportunities for entrepreneurship. Transport infrastructure is threatened by heat waves, landslides, and flooding, while electric vehicles, public transport, and circular economy solutions for batteries are poised to improve air quality in urban centers and reduce transport GHG emissions. Water resources are strained by glaciers melting, altered river flows, and increased flood and water stress risks. However, opportunities exist to enhance water efficiency and capture and use rainwater. Urban development will suffer from heat stress and stormwater flooding, affecting vulnerable groups. Urban greening offers the possibility of improved livability, as does reducing GHG emissions from buildings by improving insulation, heat efficiency, and passive cooling. Public health is challenged by increased exposure to heat stress, flooding, malnutrition, and communicable diseases. Disaster risk management must prepare for more frequent and severe weather events that impose fiscal pressures. Engineered and nature-based solutions will be key, along with enhancements to early warning systems. The private sector and financial institutions will grapple with physical and transitional risks due to shifts in technology, markets, and regulations. However, green lending portfolios and sustainability-linked finance create new opportunities for domestic and international private investment. DMCs must implement policies to ensure a just transition for all population groups toward a resilient and low-carbon economy. The costs of inaction far outweigh the costs of action, and opportunities for sustainable green growth should be pursued across CWA DMCs without delay. Further details of the impacts of climate risks and opportunities across key sectors are explained in Appendix 1, along with data demonstrating the scale of the sectors across each DMC.

Country Climate Priorities

20. DMCs have different climate finance needs, and while adaptation finance is generally a priority for all DMCs, some DMCs stand out for their mitigation finance needs. Their risk of missing 2030 NDC targets, as projected based on GHG emissions from 2017 to 2021 and their ND-GAIN adaptation gap ratings, requires an assessment identifying which DMCs need to weigh investment toward climate mitigation, toward adaptation, or toward a balance between the two (footnote 6) (Table 6). Pakistan and Turkmenistan are top priorities for climate finance due to relatively high absolute GHG emissions

and large adaptation gaps. Uzbekistan follows in both mitigation and adaptation needs. Kazakhstan is a high priority for mitigation but lower for adaptation, though some regions have high adaptation needs, while Tajikistan and Afghanistan are the opposite, with high adaptation needs but lower mitigation priorities. Based on the data presented in Appendix 1 on the impacts of climate risks across sectors, sector priorities have been identified for each DMC according to the scale of need within the sector and some additional considerations explained in Table 6. Generally, energy, transport, agriculture, and urban development are key sectors requiring high mitigation finance, while agriculture, water, social infrastructure, and disaster risk management are key sectors requiring adaptation finance.

Table 6: Assessment of ADB Developing Member Countries' Climate Priorities

Country	Assessment of Climate Priority for the Country[a]	Rationale for Determining Climate Priority for the Country	Sectors Identified Requiring Mitigation Action at Scale[b]	Sectors Identified Requiring Adaptation Action at Scale[c]
Afghanistan	A > M	Low GHG emissions; relatively low risk of missing NDC target; very high adaptation gap	Energy (#), agriculture (#)	Agriculture and water (=), social sectors (~), disaster risk management (~), urban development (~)
Armenia	A > M	Relatively low GHG emissions; relatively low risk of missing NDC target; medium adaptation gap	Energy (#), transport (#), urban development and housing (#)	Agriculture and water (~), social sectors (~), urban development (~), disaster risk management (~), transport (~)
Azerbaijan	A = M	Medium GHG emissions; medium risk of missing NDC target; medium adaptation gap	Energy (+), urban development and housing (#), transport (#), agriculture (#), fugitive emissions (#), private sector and finance ($)	Social sectors (~), agriculture and water (~)
Georgia	A = M	Relatively low GHG emissions; relatively low risk of missing NDC target; low adaptation gap	Energy (#), transport (#), waste sector (#), urban development and housing (#), private sector and finance ($)	Agriculture and water (~), social sectors (~), urban development (~), disaster risk management (~), transport (~)
Kazakhstan	M > A	Very high GHG emissions; high risk of missing NDC target; low adaptation gap	Energy (+), transport (+), agriculture (+), urban development and housing (+), private sector and finance (+), fugitive emissions (+)	Agriculture and water (~), social sectors (~). disaster risk management (~), transport and energy infrastructure (=)
Kyrgyz Republic	A = M	Relatively low GHG emissions; high risk of missing NDC target; medium adaptation gap	Energy (#), agriculture (#), urban development and housing (#), transport (#)	Agriculture and water (~), social sectors (~), disaster risk management (~), urban development (~)
Pakistan	A > M	Very high GHG emissions; low risk of missing NDC target; very high adaptation gap	Energy (+), agriculture (+), manufacturing and construction (+), urban development and housing (+), transport (+), private sector and finance ($)	Agriculture and water (~), urban development (~), transport and energy infrastructure (=), social sectors (~), disaster risk management (~)

continued on next page

Table 6 continued

Country	Assessment of Climate Priority for the Country[a]	Rationale for Determining Climate Priority for the Country	Sectors Identified Requiring Mitigation Action at Scale[b]	Sectors Identified Requiring Adaptation Action at Scale[c]
Tajikistan	A > M	Relatively low GHG emissions; medium risk of missing NDC target; medium–high adaptation gap	Energy (#), agriculture (#), industrial processes (#), urban development and housing (#)	Agriculture and water (=), social sectors (~), disaster risk management (~), urban development (~)
Turkmenistan	A = M	Medium GHG emissions; high risk of missing NDC target; high adaptation gap	Energy (+), transport (+), fugitive emissions (+)	Agriculture and water (~), disaster risk management (~), social sectors (~)
Uzbekistan	A = M	Medium GHG emissions; high risk of missing NDC target; medium adaptation gap.	Energy (+), transport (+), urban development and housing (+), agriculture (+), private sector and finance ($)	Agriculture and water (~), transport and energy infrastructure (=), social sectors (~), disaster risk management (~)

A = adaptation, GHG = greenhouse gas, M = mitigation, NDC = nationally determined contribution.

[a] Adaptation greater than, equal to, or less than mitigation.

[b] Actions are needed in all sectors and countries. (+) = GHG emissions for this sector are high in absolute terms compared with other Central and West Asia (CWA) countries. (#) = GHG emissions for this sector are high for this country compared with other sectors. ($) = Private investment can be readily leveraged for mitigation action using, for example, renewable energy.

[c] Actions are needed in all sectors and countries. (=) = The scale of this sector in this country, compared with other CWA countries, implies a high need for adaptation interventions. (~) = The adaptation needs of this sector in this country, compared with other CWA countries, are recognized as high by the country.

Source: Asian Development Bank.

Regional Cooperation Needs for Climate-Resilient and Low-Carbon Development

21. National climate action should be complemented with regional interventions for effective climate action. Table 7 is an assessment of priorities for regional cooperation for climate resilience. There are opportunities to deliver multiple benefits to climate adaptation and GHG reduction, environmental sustainability, and social and economic development. The overarching climate priority for regional cooperation is for each DMC to play its part in reducing GHG emissions in line with its NDC and support neighboring countries' efforts to adapt to climate change, as outlined in NAPs.

Table 7: Assessment of Regional Cooperation Priorities for Climate Resilience

Sector	Regional Cooperation Priority	Climate Benefits
Climate policies, public sector, and governance	• Capacity building to promote climate policy frameworks aligned with the Paris Agreement • Promoting an enhanced transparency framework for measurement, reporting, and verification as per Paris Agreement Article 13 • Integrating DMCs into international carbon markets as per Paris Agreement Article 6 • Mobilizing private climate finance and enhanced access to green capital markets as per Paris Agreement Article 2 • Enhancing regulation and the legislative environment for public–private partnership • Promoting carbon pricing and carbon taxation • Supporting for the revision of fossil fuel subsidy systems and a fair transition to realistic utility tariffs • Assessing climate risks to socioeconomic sectors (i.e., agriculture, energy, health, transport, urban, and water) and developing innovative risk management and transfer solutions • Connecting and enhancing social protection systems and disaster risk management systems	Enhanced adaptation outcomes; opportunities to reduce GHG emissions and sequester carbon
Water	• Regional initiatives on water security • Cross-border and catchment-wide sustainable water management that is integrated with the need for drinking water supply, the environment, agriculture, and energy; includes cooperation on glacial melt resources, river basins, groundwater, and salinity, such as water productivity and water accounting with the use of satellite and drone technologies	Enhanced adaptation outcomes; opportunities to reduce GHG emissions and sequester carbon
Agriculture and forestry	• Regional initiatives on food security, sustainable agricultural practices, and sustainable land management • Regional initiative to develop agricultural climate information systems, benefiting farmers through the provision of digitally processed data and extension services • Biosecurity to minimize pest and disease impacts on agriculture	Enhanced adaptation outcomes; opportunities to reduce GHG emissions and sequester carbon
Energy	• Regional initiatives on energy security, including enabling renewables • Resilient supply and transmission of energy across borders • Safe operation of such energy systems as nuclear power	Enhanced adaptation outcomes; opportunities to reduce GHG emissions
Transport and communications	• Resilient and sustainable regional road, rail, and port transport links for mobility and critical supply chains • Resilient regional communications networks	Enhanced adaptation outcomes; opportunities to reduce GHG emissions
Health	• Prevention and management of epidemics and other communicable diseases • Regional initiatives to tackle air and water pollution • Regional initiatives to reduce health impacts from heat and climate extremes	Enhanced adaptation outcomes; opportunities to reduce GHG emissions
Environment	• Regional initiatives to enhance biodiversity and conserve the natural environment, including for cross-border conservation areas • Regional initiatives to prevent and combat desertification	Enhanced adaptation outcomes; opportunities to reduce GHG emissions and sequester carbon
Trade and economy	• Enabling cross-border trade in goods, services, finance, and tourism • Enabling potential for residing and working in neighboring countries	Enhanced adaptation outcomes; opportunities to reduce GHG emissions

continued on next page

Table 7 continued

Sector	Regional Cooperation Priority	Climate Benefits
Waste	• Regional initiatives to safely and sustainably manage waste and prevent pollution of land, water, and air	Enhanced adaptation outcomes; opportunities to reduce GHG emissions
Disaster risk reduction	• Cross-border and catchment-wide initiatives to prevent and manage flooding • Connection and enhancement of social protection systems, early warning systems, and disaster risk management systems	Enhanced adaptation outcomes; opportunities to reduce GHG emissions and sequester carbon
Security	• Regional efforts to combat preventable cross-border economic migration and manage asylum seekers • Regional efforts to promote peace and prevent conflicts	Enhanced adaptation outcomes; opportunities to reduce GHG emissions

DMC = developing member country, GHG = greenhouse gas.

Sources: CAREC 2030 Strategic Framework; country nationally determined contributions and national action plans; and engagement with DMCs and experts across the Asian Development Bank.

22. The Central Asia Regional Economic Cooperation (CAREC) program aims to enhance climate policies and action through regional collaboration in 11 countries.[9] The CAREC Climate Change Vision, endorsed in November 2023, provides a strategic framework to coordinate and accelerate regional climate initiatives.[10] It emphasizes sharing best practices, identifying priority sectors for adaptation and mitigation, mobilizing climate finance, and advocating for a unified CAREC voice on the global climate stage. To implement this vision, the Working Group on Climate Change was established to develop the CAREC Climate Change Action Plan (forthcoming), which intends to integrate with actions identified in ADB's climate action road map for CWA and ADB's country-specific CAPs. The CAREC Climate Change Action Plan (forthcoming) will prioritize regional climate adaptation and mitigation projects, enhance coordination among development partners, and optimize resources for regional climate actions. CAREC will support capacity building and help to fund climate action in CWA.[11] To guide these efforts, CAREC will establish the high-level Climate Change Steering Committee, supported by the Working Group on Climate Change. The CAREC Institute will enhance knowledge and capacity building, while the CAREC Climate and Sustainability Project Preparatory Fund will support the preparation of bankable regional projects focused on climate and sustainability objectives.[12] Similarly, the CAREC Water Pillar provides an effective platform to discuss a regional approach in tackling transboundary issues (i.e., water scarcity) and identify investment programs and projects with regional and national significance.

9 These 11 countries are Afghanistan, Azerbaijan, the People's Republic of China, Georgia, Kazakhstan, the Kyrgyz Republic, Mongolia, Pakistan, Tajikistan, Turkmenistan, and Uzbekistan. ADB placed its regular assistance to Afghanistan on hold effective 15 August 2021.

10 ADB and CAREC. 2024. *Regional Action on Climate Change: A Vision for CAREC*.

11 Key intervention areas include (i) climate risk, preparedness, and health, focusing on regional infrastructure, health facilities, early warning systems, disaster risk management, and financing; (ii) the water–energy–food security nexus; (iii) low-carbon growth, including decarbonization of transport corridors, trade, and regional grid integration of renewable resources; and (iv) the CAREC Climate Platform, concentrating on financing, capacity development, and climate negotiations.

12 This multi-donor trust fund facility is administered by ADB using contributions from development partners. The fund will help build the capacity of CAREC countries to design and implement regional bankable projects focused on climate and sustainability objectives.

23. Other regional strategies and plans complement this road map in guiding collaboration and prioritizing climate action in the region. The Regional Climate Change Adaptation Strategy in Central Asia was developed in 2021 by the five Central Asian countries—Kazakhstan, the Kyrgyz Republic, Tajikistan, Turkmenistan, and Uzbekistan—with support from the Green Central Asia Initiative of the Government of Germany.[13] The strategy identifies several priority actions for climate adaptation: (i) strengthening regional coordination for climate change adaptation;[14] (ii) creating mechanisms to develop and implement adaptation projects and programs and attract financing;[15] (iii) improving adaptive capacity through the accumulation and sharing of knowledge and scientific cooperation;[16] and (iv) developing climate monitoring, information exchange, and forecasting systems.[17] The strategy emphasizes the importance of regional cooperation to optimize the use of national potential and ensure joint activities, particularly in water and energy resources and ecosystem management.

Climate Finance Provided by Multilateral Development Banks and International Finance Institutions in Central and West Asia

24. Multilateral development banks (MDBs) and other international finance institutions (IFIs) have made significant climate finance investments in CWA, but some countries and sectors are falling behind. From 2018 to 2021, MDBs and IFIs invested $21.6 billion across the 10 countries in CWA (Figure 4). From the MDBs, World Bank Group (WBG) provided 34%, the European Bank for Reconstruction and Development (EBRD) 15%, ADB 14%, and the Asian Infrastructure Investment Bank (AIIB) 3%. Also significant was climate finance provided by France at 7% of the total, Germany at 7%, and Japan at 8%. Pakistan and Uzbekistan each received almost 30% of this climate finance, with the other countries receiving much smaller portions, from less than 1% to 13% (Figure 5). The WBG provided over 60% of the climate finance to Pakistan, while climate finance to Uzbekistan was more diverse, with several MDBs and IFIs involved. The energy sector was allocated the largest share of climate finance, at 37% (Figure 6).[18] The WBG, the EBRD, Japan, and ADB all provided significant amounts of climate finance to this sector, with the WBG providing the most during the 2018–2021 period. Climate finance investments in transportation and storage were similar across the WBG, the

[13] Deutsche Gesellschaft für Internationale Zusammenarbeit (GIZ). 2021. *Regional Climate Change Adaptation Strategy for Central Asia and Regional Action Plan For a Joint Political Dialogue on Climate, Environment and Security*.

[14] Such as establishing and supporting regional mechanisms for coordinating joint activities; building the capacity of national and regional authorities to manage climate change adaptation processes based on best practices and advanced experience; and harmonizing policies, plans, laws, and regulations related to vulnerability assessment and climate change adaptation between countries.

[15] Such as identifying and utilizing joint opportunities for developing adaptation projects; mobilizing resources from various sources, including climate mechanisms, development partners, international banks, the private sector, and national budgets; and supporting the participation of Central Asian countries in international climate negotiations and processes.

[16] Such as enhancing climate monitoring networks and joint capacity building for climate change impact and forecast assessment; developing and sharing training materials and methodologies for climate change adaptation; and promoting regional mechanisms for knowledge accumulation and sharing, such as the Informational Portal on Climate Change in Central Asia and the Climate Transparency Center.

[17] Such as expanding climate observation networks and introducing advanced technologies for climate monitoring; developing regional climate models and conducting joint glaciological monitoring; and creating systems for calculating threshold values of hydrometeorological emergencies.

[18] Organisation for Economic Co-operation and Development. *Climate-Related Development Finance Portal* (accessed 8 August 2024). Sources of climate financing are public (multilateral, bilateral, and national development finance institutions) and private finance.

EBRD, and ADB. Several DMCs and sectors have been underfinanced toward meeting their climate needs, and various financing instruments have been underutilized. However, a large reason for this is little appetite or capacity in DMCs to borrow from MDBs and IFIs for investment in climate action. Appendix 2 provides an overview of strategic frameworks, climate finance targets, financing modalities, and pricing approaches of selected MDBs and IFIs active in CWA.

25. Adaptation finance investment levels in CWA by MDBs and IFIs has been well below those for mitigation finance. From 2018 to 2021, more than two-thirds of the finance provided was for climate mitigation. The WBG provided in absolute terms the largest amount of finance for adaptation (40%), while the EBRD (14%), ADB (13%), and the AIIB (3%) all provided far smaller amounts. The institutions most favorable to adaptation finance over mitigation in their regional climate finance portfolios were EU institutions not including the European Investment Bank (EIB) at 59% and the Green Climate Fund (GCF) at 50%. Adaptation proportions were similar for bilateral development assistance from Germany at 38% and France at 26%, while 84% of Japan's climate finance was for mitigation. Given the severe impact of climate change in CWA, accelerating investment in adaptation and resilience is critical. The agriculture, forestry, and fishing sector and the banking and financial services sector received relatively little climate finance from ADB or other MDBs.

26. The French development bank Agence Française de Développement (AFD) is the only other major IFI that has developed a regional strategy for CWA, but it is not specifically focused on climate action, and AFD does not have a current climate financing target. AFD's Eurasia Regional Strategy outlines four priorities: (i) resilience and urban adaptation, (ii) integrated and sustainable territorial management and conservation of natural resources, (iii) support for energy transition, and (iv) sustainable and inclusive economic and political models. The WBG is aiming for 35% of its project finance over a 5-year period to be climate finance, half of it for adaptation. The WBG developed country climate and development reports for four countries in the region—Azerbaijan, Kazakhstan, Pakistan, and Uzbekistan—and has a South Asia strategy that includes Pakistan.[19] The EBRD, AIIB, and EIB aim to have more than 50% of their annual investments to be in green or climate finance. The EBRD developed green city action plans for selected cities in CWA. Its priority thematic areas for Central Asia are natural capital, sustainable connectivity, and sustainable food systems.[20] The AIIB and EIB have no specific strategies for CWA that guide their investment in the region. This road map offers a regional perspective for delivering climate action and sustainable impact.

[19] WBG. 2024. *Country Climate and Development Reports.*

[20] EBRD. What is the EBRD's Green Economy Transition approach?

Figure 4: Climate Finance Mobilized by Selected Institutions in Central and West Asia, 2018–2021
($ million)

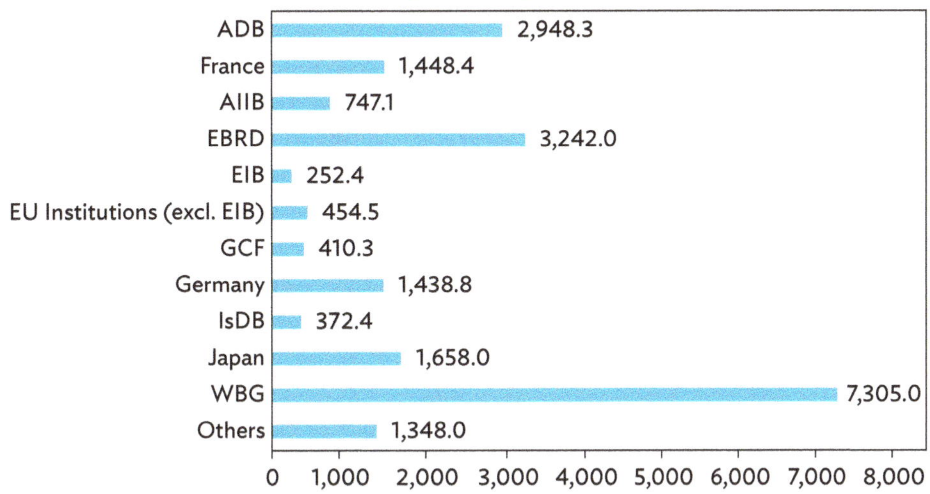

ADB = Asian Development Bank, AIIB = Asian Infrastructure Investment Bank, EBRD = European Bank for Reconstruction and Development, EIB = European Investment Bank, EU = European Union, GCF = Green Climate Fund, IsDB = Islamic Development Bank, WBG = World Bank Group.

Note: The figure shows multilateral development banks that together make up 95% of the climate finance allocated in 2018–2021, with lesser contributors omitted.

Source: Organisation for Economic Co-operation and Development. 2024. Climate-Related Development Finance Portal (accessed 1 October 2024).

Figure 5: Climate Finance Mobilized in Central and West Asia, by Country and Selected Institutions, 2018–2021

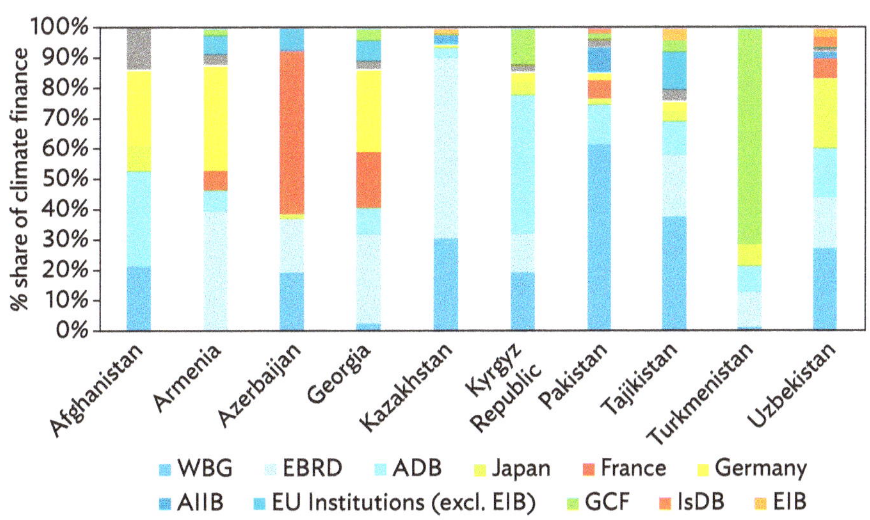

ADB = Asian Development Bank, AIIB = Asian Infrastructure Investment Bank, EBRD = European Bank for Reconstruction and Development, EIB = European Investment Bank, EU = European Union, GCF = Green Climate Fund, IsDB = Islamic Development Bank, WBG = World Bank Group.

Note: The figure shows multinational development banks that together make up 95% of the climate finance allocated to Central and West Asia in 2018–2021, with lesser contributors omitted.

Source: Organisation for Economic Co-operation and Development. Climate-Related Development Finance Portal (accessed 6 October 2024).

Figure 6: Climate Finance Mobilized in Central and West Asia, by Sector and Selected Institutions, 2018–2021

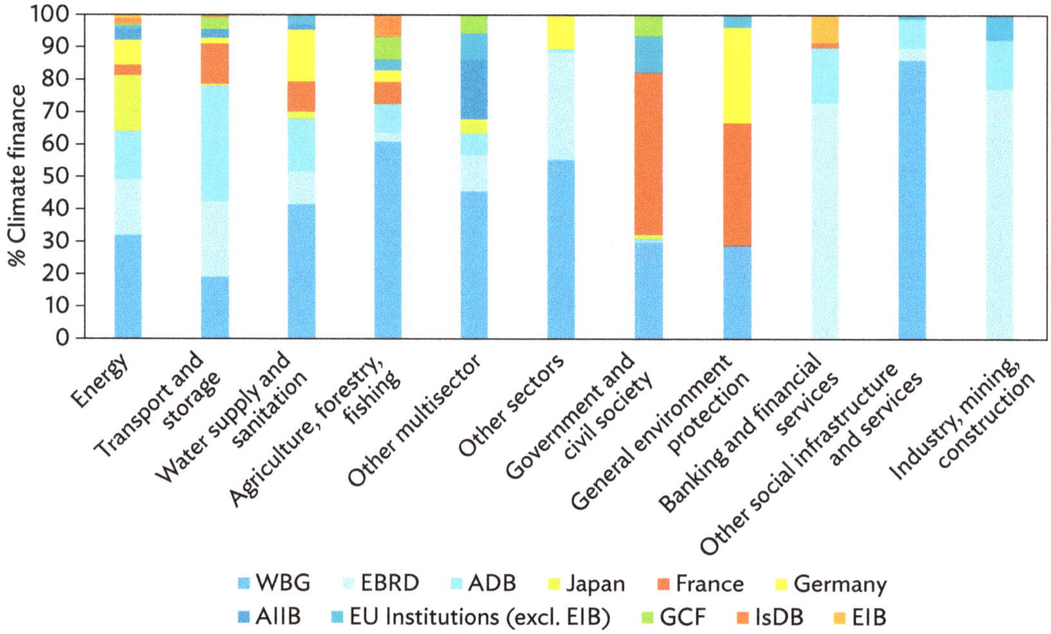

ADB = Asian Development Bank, AIIB = Asian Infrastructure Investment Bank, EBRD = European Bank for Reconstruction and Development, EIB = European Investment Bank, GCF = Green Climate Fund, IsDB = Islamic Development Bank, WBG = World Bank Group.

Note: The figure shows sectors that together make up 95% of the climate finance allocated in 2018–2021, with lesser sectors omitted.

Source: Organisation for Economic Co-operation and Development. Climate-Related Development Finance Portal (accessed 1 October 2024).

Climate Finance Needs in Central and West Asia

Climate Finance Gap

27. Recent studies indicate that developing countries need to spend the equivalent of 4%–6.5% of GDP annually for climate mitigation and adaptation.[21] For CWA, the annual climate finance requirement is estimated at up to $100 billion. International climate finance to CWA needs to substantially accelerate to meet NDC requirements of CWRD DMCs.[22]

[21] M.S. Ahluwalia and U. Patel. 2022. Climate Change Policy for Developing Countries. In H. Kohli et al., eds. *Envisioning 2060*. Penguin Random House; ADB. 2023. *CAREC 2030—Supporting Regional Actions to Address Climate Change. A Scoping Study*; Climate Policy Initiative. 2024. Global Landscape of Climate Finance 2024: Insights for COP29. October; and International Monetary Fund. 2024. Unlocking Climate Finance in Asia-Pacific: Transitioning to a Sustainable Future. *Departmental Paper DP/2024/001*. Asia and Pacific and Statistics Departments. January.

[22] ADB. 2023. Climate Finance Landscape of Asia and the Pacific.

28. Private climate finance amounted to $168.1 billion in Asia and the Pacific in 2018 and 2019, but almost all of it was for mitigation. Private finance is 32% of the total tracked climate finance (footnote 24). Of this private finance, 99% was allocated to mitigation efforts. The Climate Policy Initiative's Global Landscape report identified key private stakeholders, including corporate entities, households, commercial financial institutions, and institutional investors, with a primary focus on climate-related infrastructure and, to a lesser extent, sustainable agriculture and land use.[23] Across Asia and the Pacific, corporations were the largest contributors, providing $72.3 billion, followed by households, which invested $47.8 billion mainly in residential photovoltaic systems and electric vehicles. Commercial financial institutions doubled their contributions from 2018 to 2019, when they invested $45.1 billion primarily in clean transport and renewable energy. Private funds allocated $1.5 billion, and institutional investors $1.3 billion, with a focus on energy. However, only 1% of private finance was directed toward adaptation finance, highlighting the perception of inadequate market development and limited scalability in adaptation projects.

29. Some DMCs are more able than others to attract and encourage private climate finance, and this is an area that needs addressing across DMCs. Georgia and Pakistan are notable for facilitating private finance investment through supportive regulation. Private finance investment in these two countries spanned renewable energy, low-carbon transport, and energy-efficient buildings, underscoring its pivotal role in climate change mitigation. However, private finance had little involvement in adaptation measures despite the growing need for climate resilience. Barriers to greater private climate finance involvement include high perceived risks, difficulty in reliably quantifying benefits, unclear policy signals, and insufficient incentives for investment in emerging technologies and markets. Additionally, the complexity of tracking and reporting adaptation finance, coupled with the lack of standardized metrics, further impedes private adaptation finance. Enhanced private sector engagement requires appropriate regulatory norms, concessional financing instruments, and capital market reform.

ADB's Climate Finance Portfolio

30. ADB's climate finance from its own resources for sovereign and nonsovereign operations in CWA from 2019 to 2023 amounted to $4.9 billion.[24] Climate finance comprised 35% of total commitments to CWA in the last 5 years.[25] Approximately 43% of climate finance contributed to adaptation, and 57% to mitigation. Total nonsovereign climate finance to DMCs in CWA encompassed 10% of total climate finance, almost exclusively focused on mitigation.[26] Pakistan and Uzbekistan together constitute more than 77% of ADB's total climate finance delivered to CWA DMCs (Figure 7). ADB had no operations in Turkmenistan during these 5 years, while Azerbaijan and Kazakhstan had mostly private sector operations. The sector portfolio across CWA DMCs is diverse but dominated by infrastructure (Figure 8).

[23] Climate Policy Initiative. 2023. *Global Landscape of Climate Finance 2023.*

[24] ADB Climate Finance Database (internal) (accessed August 2024). Data do not include financial contributions through regional projects and technical assistance, which cover several countries.

[25] Total commitments in CWA from 2019 to 2023: $13.2 billion.

[26] A total of 97% of nonsovereign funds were invested for mitigation objectives.

Figure 7: ADB Total Sovereign and Nonsovereign Climate Finance in Central and West Asia by Country, 2019–2023
($ million)

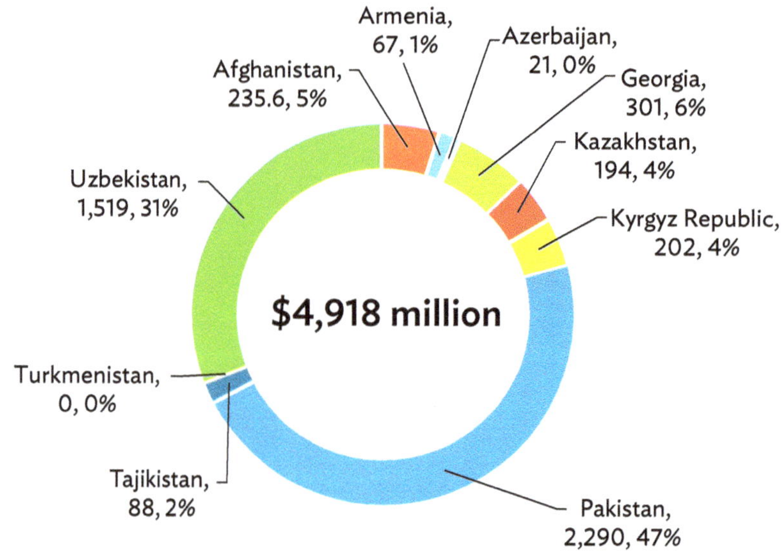

Source: Asian Development Bank. Climate Finance Database (internal) (accessed 14 November 2024)

Figure 8: ADB's Sovereign and Nonsovereign Climate Finance in Central and West Asia by Country and Sector, 2019–2023

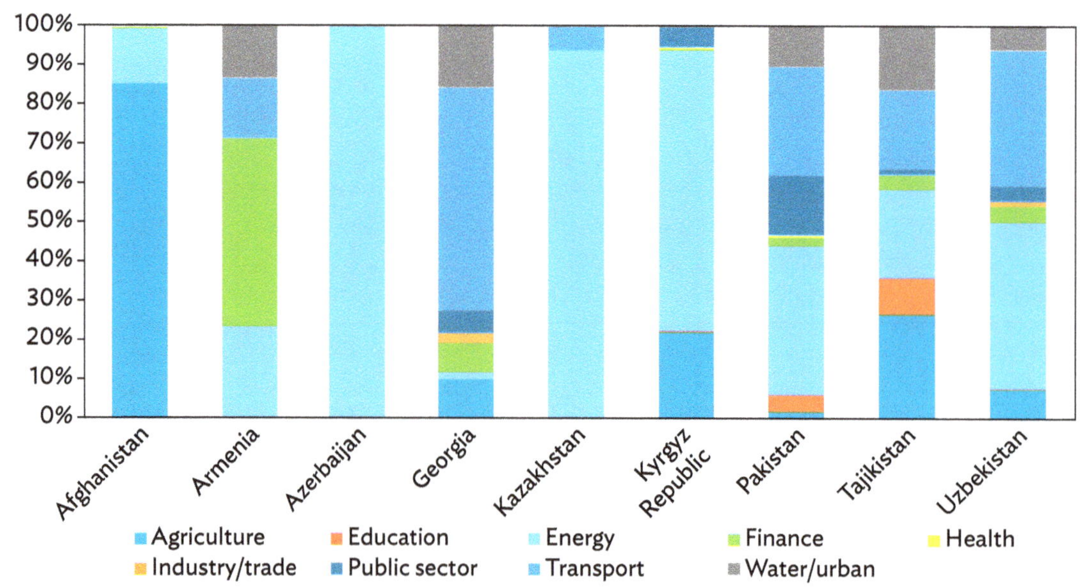

Source: Asian Development Bank. 2023. *Climate Finance Landscape for Asia and the Pacific*.

31. ADB operations in CWA aim to expand and diversify across sectors and grow investment in countries with large climate finance needs to support the transition. Infrastructure sectors such as transport and energy still dominate DMC country portfolios. However, sectors such as education, health, and finance are increasingly broadening their shares (Figures 9 and 10, Appendix 2).

Figure 9: Annual ADB Climate Finance in Central and West Asia, 2019–2023
($ million)

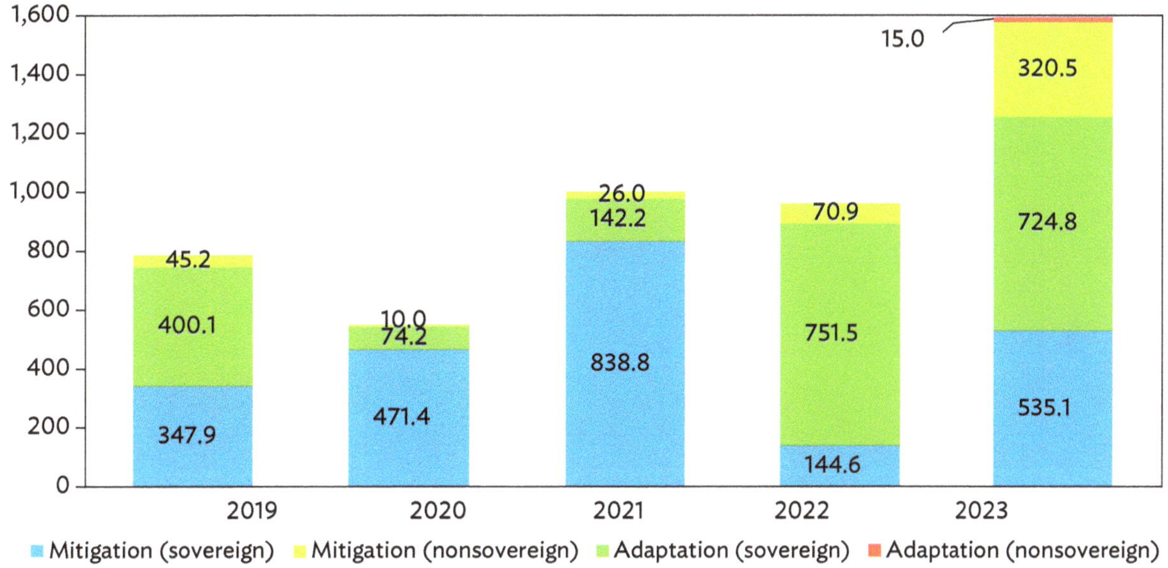

ADB = Asian Development Bank.
Source: ADB. Climate Finance Database (internal) (accessed 26 February 2024).

Figure 10: ADB Climate Finance in Central and West Asia by Sector, 2019–2023

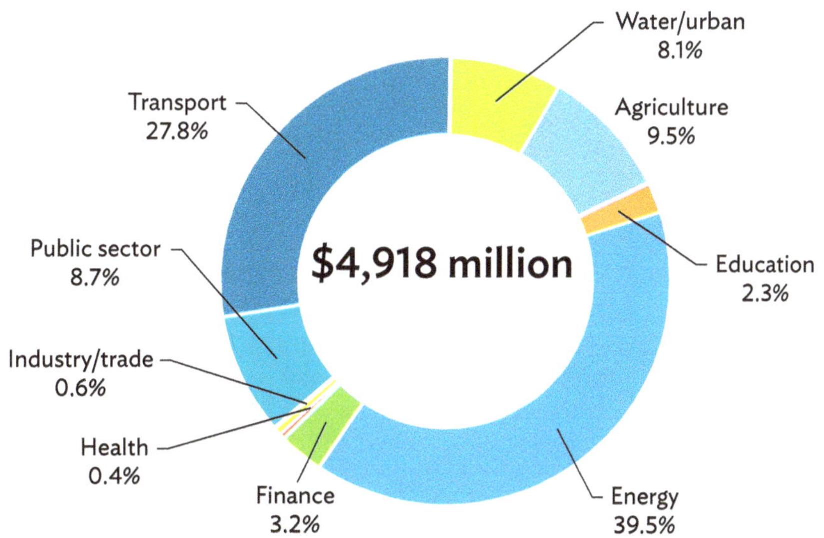

Source: Asian Development Bank. Climate Finance Database (internal) (accessed 26 February 2024).

32. ADB mobilizes cofinancing to support climate action in CWA. From 2019 to 2023, 15% of the ADB Central and West Asia Department projects were cofinancing partnerships (22 projects with 12 cofinanciers). The majority of cofinancing was in energy, transport, and public sector management, mostly in Pakistan and Uzbekistan. Cofinancing combines ADB's own resources with resources from other development partners, leveraging larger investments for the benefit of DMCs. Cofinancing allows risks to be shared across development partners, concessional finance to be blended with market rate finance, and the various development partners to contribute expertise and experience according to their comparative advantages. Cofinancing partners of projects in CWA during recent years were the AFD, AIIB, EBRD, and WBG (Figure 11). ADB can increase the climate impact delivered by increasing cofinancing with existing and new partners, and by aligning with partners' priorities and investment approaches (e.g., approaches in countries, sectors, and investment types and modalities). Alternative cofinancing options should continue to be enhanced through trusted partnerships with global climate funds such as the GCF and the Global Environment Facility, significantly leveraging additional means for ADB operations in DMCs. Lessons can be learned from current and emerging projects in partnership with these funds. As administrative requirements, timelines, and the specifics of project preparation can be barriers to cofinancing, solutions need to be developed to alleviate these potential obstructions. Parallel financing is another way in which climate impact can be delivered with partners by, for example, financing separate but complementary projects.

Figure 11: Cofinancing Contributions to ADB Sovereign Climate Finance in Central and West Asia, 2019–2023

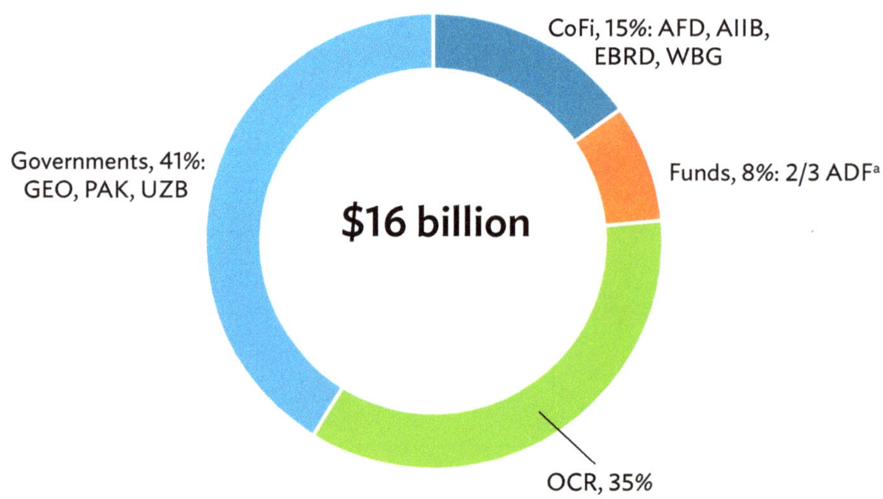

AFD = Agence Française de Développement, AIIB = Asian Infrastructure Investment Bank, ADF = Asian Development Fund, CoFi = Cofinancing, EBRD = European Bank for Reconstruction and Development, GEO = Georgia, OCR = ordinary capital resources, PAK = Pakistan, UZB = Uzbekistan, WBG = World Bank Group.

[a] 90% from ADF, the rest from Japan Fund for Prosperous and Resilient Asia and the Pacific, Urban Climate Change Resilience Trust Fund, Education Above All Foundation, High Level Technology Fund, OPEC (Organization of the Petroleum Exporting Countries) Fund, and Green Climate Fund.

Source: Asian Development Bank. Climate Finance Database (internal) (accessed 26 February 2024).

3 STRATEGIC DIRECTIONS (THE ROAD MAP)

33. ADB is the climate bank of Asia and the Pacific. This *Climate Action Road Map for Central and West Asia, 2025–2030* aims to shift the bank's operations to become more responsive to DMCs' climate action needs and priorities. In alignment with its climate change action plan and new operating model, ADB's operations in the region will help DMCs deliver climate-responsive economic and social development through high-quality sovereign and nonsovereign operations.[27]

How ADB Engages in Central and West Asia

34. ADB will deploy the One ADB approach to enhance collaboration across sectors, groups, and teams and thus deliver integrated multisector solutions that are systematic and resource efficient. This approach promotes collaborative ways and aims to identify solutions tailored to client needs, making the road map client centric. It prioritizes the climate-related needs of CWA DMCs and benefits to the people at the core of its support. Additionally, the road map recognizes the need to mobilize additional climate finance, achieved through expanding cofinancing and partnerships, and through mobilizing private sector investments in climate solutions and innovations. A key precondition for this is a conducive business environment that supports investment in climate and green finance. ADB will continue to strengthen its support in CWA at three levels:

Upstream: Strategic Engagement for Enhanced Policy Frameworks

35. ADB is intensifying its efforts to support DMCs in CWA in their climate change initiatives by integrating climate action into national and regional policy frameworks through country partnership strategies. ADB aims to offer ways to translate low-carbon transitions as economic growth opportunities rather than trade-offs. This involves embedding climate considerations into country partnership strategies and enhancing staff capacity to lead climate dialogues. ADB will promote improved policy-based loans to reinforce DMCs' policy frameworks and provide technical assistance and capacity building to align national contributions and adaptation plans with development policies. Additionally, ADB will continue to foster private sector investment in climate initiatives and drive regional cooperation to address climate challenges.[28] A just and inclusive transition to low-carbon, climate-resilient economies will be a focus, ensuring social inclusion and intergenerational equity. Gender will be central to ADB's climate agenda, empowering women through climate policies. Finally, ADB will incorporate climate diagnostics into country programming and thus leverage its convening power to develop long-term solutions for DMCs.

---

27 ADB. 2022. *Organizational Review. A New Operating Model to Accelerate ADB's Transformation Toward Strategy 2030 and Beyond.*

28 This will be done through the institutional structures of the CAREC Program and, under it: ADB. 2023. *Supporting Regional Actions to Address Climate Change.*

Midstream: Embedding Climate Action in Core Institutions and National Systems

36. ADB is committed to integrating climate action into DMCs' institutional systems. This includes embedding climate priorities in budgeting and budget-tracking processes, public procurement, and national fiscal frameworks. ADB will continue to enhance DMCs' access to climate finance by improving domestic resource mobilization, implementing carbon pricing, and leveraging private sector finance through public–private partnerships (PPPs) and blended finance, including through partnerships with other development partner organizations, donors, and funds. Additionally, ADB aims to utilize risk transfer mechanisms to protect fiscal balances against climate risks and support investments in resilience through such innovative financial instruments as green and blue bonds.[29] ADB will also help DMCs align their sector policies with national climate plans and strengthen infrastructure governance. To support these initiatives, ADB will continue to work on country diagnostics and sector assessments, invest to refine economic analyses related to climate change, develop climate-focused fiscal policies, and increase the bankability of climate-resilient projects to attract private investment.

Downstream: High-Quality Operations and Implementation

37. ADB is intensifying its efforts to expand the scale and enhance the quality of operations in CWA, focusing on such key sectors as energy, transport, and water infrastructure, which are crucial for decarbonization and adaptation. ADB is committed to providing upstream support to both sovereign and nonsovereign operations to bolster decarbonization opportunities and manage climate-related risks. Additionally, ADB is working to enhance the finance sector's role in developing green climate finance products and improving disclosure and reporting standards. ADB's focus on sustainable, climate-resilient infrastructure will ensure long-term asset sustainability through efficient operation and maintenance, as well as comprehensive training and capacity building. Appendix 3 provides a few examples showing the interconnectedness of ADB upstream, midstream, and downstream interventions.

38. Supporting DMCs' climate mitigation and adaptation policies requires transformative and systemic changes at the policy, economy, and social level. However, integrating climate aspects into sustainable and inclusive development plans is still rather new and often challenging for DMCs in CWA, requiring upstream support from development partners. To increase the effectiveness and volume of climate finance, interventions need to shift from supporting incremental change to fostering transformational change.[30] Thus, identifying climate finance approaches—especially results-based loans, policy-based loans, sector development programs—with DMCs is imperative for a climate bank.

---

[29] Corporate bonds are debt securities issued by corporations to raise capital for various purposes, such as expanding operations, funding new projects, or refinancing existing debt. Corporate green bonds are a specific type of corporate bond where the proceeds are exclusively used to finance or refinance projects that have a positive environmental impact. Corporate blue bonds are similar to green bonds but are specifically designed to finance projects related to the conservation and protection of marine ecosystems.

[30] Incremental change is achieved through climate activity that is additional to the intervention's main activities but, on its own, does not create deep, systemic, or sustainable change. Transformational change, on the other hand, arises through a project or larger program that pursues deep, systemic, and sustainable change, with potential to seriously address climate change. See ADB. 2022. *ADB Climate Change Operations Roadmap in South Asia (2023–2030)* (internal).

Strategic Directions for Climate Action

39. **Climate action needs and opportunities in CWA from 2025 to 2030.**[31] Proposed interventions are based on analysis of DMCs' key climate policy documents (e.g., NDCs, long-term low-emission development strategies, and sector plans) and extensive consultations with DMCs, ADB, other development partners, and key stakeholders. The climate actions identified align with ADB Strategy 2030 sector directional guides. The identified climate finance solutions supported by ADB in CWA are not an exhaustive list. The interventions will be regularly reviewed and updated based on ADB's growing portfolio and diversified investments. The list of actions is intended to become a guide to operations across DMCs in CWA. Strategic directions for ADB's support and investment in CWA are structured across four essential thematic areas: governance, economy, society, and environment. ADB's seven sectors are allocated into these four areas. ADB acknowledges that climate-change-oriented development solutions require approaches that cut across sector boundaries and the four essential thematic areas.

40. **Thematic area 1: governance and public sector reform.** ADB will help DMCs align their climate policies with the Paris Agreement. This includes updating NDCs and developing national adaptation plans (NAPs) and sector-specific strategies. ADB plans to integrate climate policies into broader economic frameworks, enhance public financial management, mobilize private investment in climate initiatives, and establish standards for low-emission and resilient infrastructure across various sectors.

41. **Thematic area 2: green and resilient economic growth.** ADB aims to increase the share of renewable energy, enhance the resilience of infrastructure under climate impacts, and support decarbonization in major industries. Initiatives include developing renewable energy systems like solar and wind, climate-proofing physical infrastructure, and promoting electric mobility and multimodal transport systems. ADB seeks to attract private sector investment through improved regulatory frameworks, PPP, and financial instruments like green bonds.

42. **Thematic area 3: resilient and inclusive societies.** ADB works to create resilient and inclusive urban environments by decarbonizing urban transport systems; promoting sustainable urban planning; and reforming policies to support green, livable cities. The bank also focuses on enhancing social infrastructure through climate-resilient health and educational facilities, integrating social protection with disaster risk management, and implementing comprehensive climate education programs.

43. **Thematic area 4: sustainable management of natural resources and the environment.** This area covers the sustainable management of landscapes, forests, and water resources, emphasizing transboundary water management and climate-smart agriculture. ADB supports the development of resilient agricultural value chains, natural resource protection, improved waste management, and the adoption of circular economy principles.

44. These climate actions are responsive to ADB's *Climate Change Action Plan, 2023–2030* (CCAP) and align with ADB Strategy 2030 sector directional guides. Climate actions that contribute to the implementation of the corporate CCAP list of climate actions are marked accordingly. ADB will support DMCs' achievement of adaptation through national and sector adaptation plans, and of mitigation

[31] Because ADB placed its regular assistance to Afghanistan on hold effective 15 August 2021, the following sections do not include climate actions for Afghanistan.

NDCs and long-term low-emission development strategy goals according to the theory of change outlined in Figure 1. The proposed interventions need to be further detailed in respective country-specific climate action plans.

Governance, Institutions, and Public Sector Policy Reform

45. **Empower climate governance and public sector management.** ADB will continue to help DMCs align climate policy frameworks with the Paris Agreement by updating NDCs, developing sector and national adaptation plans, and creating green economy transition strategies and long-term low-emission development strategies in collaboration with United Nations institutions (Box 1). DMCs need assistance in translating national climate policies into sector-specific, gender-sensitive action plans and budgets, such as transport transition road maps. ADB will prioritize helping government institutions harmonize climate policies across sectors, especially in Kazakhstan, the Kyrgyz Republic, Pakistan, and Tajikistan. Additionally, the integration of national climate policies into subnational plans will be strengthened, particularly in countries with decentralized governance like Pakistan. Project examples are Bridging the Gap between Climate Adaptation Planning and Financing in Armenia and Supporting Renewable Technology—Inclusive Heat Supply Legislation in Kazakhstan.[32]

Box 1: Accelerating Climate Actions in Asia and the Pacific Through NDC Advance

The Asian Development Bank (ADB) NDC Advance platform plays a crucial role in supporting the implementation of nationally determined contributions (NDCs) in developing member countries. The platform provides technical assistance, capacity building, and financial mobilization to help countries achieve their climate targets under the Paris Agreement. By providing targeted support and facilitating regional collaboration, the platform ensures that developing member countries are equipped to meet their NDC targets and contribute effectively to global climate efforts. This technical assistance platform is designed to guide the country toward achieving its climate goals effectively. In Armenia, Georgia, and Uzbekistan, ADB's support through the NDC Advance platform included developing urban climate assessments to integrate low-carbon and climate-resilient measures into urban planning and investment strategies, thereby aligning municipal development with broader climate objectives. NDC Advance is currently developing climate finance landscape assessments for Georgia, private sector mobilization road maps for Kasazkhstan, and support to development of NDC 3.0 for Uzbekistan.

Source: ADB. 2023. *NDC Advance: Accelerating Climate Actions in Asia and the Pacific*.

[32] The knowledge and support technical assistance provided the Government of Kazakhstan with legal and technical due diligence to develop a regulatory framework and incentive schemes for renewable energy source utilization by the heat supply sector and improvement of energy efficiency standards. Effective GHG reduction measures from the heating sector are proposed to the Ministry of Environment. The draft law and by-laws were submitted to the Parliament of Kazakhstan in December 2021 and approved by the President in July 2024. For further reading, see ADB. 2023. *Incorporating Climate Resilience in Urban Planning and Policy Making: Focus on Armenia, Georgia, and Uzbekistan*.

46. **Integrate climate policies into the larger fiscal framework.** ADB will support CWA DMCs by identifying climate policy bottlenecks and promoting green, gender-sensitive budget frameworks.[33] This includes low-emission procurement, sustainable climate debt management, and climate-sensitive tracking systems. In coordination with the International Monetary Fund, ADB will help DMCs implement climate-specific programs, such as climate public investment management assessments, with a view to accessing and utilizing international carbon markets to facilitate investment in renewable energy and developing carbon pricing policies relating to emissions trading and taxation (Box 2). Efforts will focus on decarbonizing high-emission sectors and encouraging private investment in renewables and carbon sinks. Also crucial is to align carbon pricing with national processes under the Paris Agreement's Enhanced Transparency Framework.

Box 2: ADB Sustainable Fiscal Management and Governance Improvement Program in the Kyrgyz Republic

The Asian Development Bank's proposed Sustainable Fiscal Management and Governance Improvement Program is a policy-based loan and technical assistance grant to help the Kyrgyz Republic improve governance and climate- and gender-responsive public financial management, strengthen governance in state-owned enterprises, strengthen tax policy and administration, and adopt anticorruption measures. Reform through the program aims to (i) strengthen the quality of public financial management, (ii) improve the governance of state-owned enterprises, (iii) strengthen tax policies and administration, and (iv) make public administration more transparent and accountable. The main climate-specific interventions of the policy-based loan in the program are to (i) support an assessment of public financial management of climate initiatives with a view to evaluating and developing an adequate policy framework that enhances the government's capability to effectively respond to climate and disaster hazards, and (ii) conduct a climate-responsive tax policy analysis to identify implicit biases that exacerbate climate-related challenges and inequality. This analysis will help develop tax policies that incentivize climate-responsive investments by the private sector.

Source: Asian Development Bank. *Kyrgyz Republic: Sustainable Fiscal Management and Governance Improvement Program (Subprogram 1)*.

47. **Strengthening the management of fiscal climate risks.** ADB will assist DMCs' active management of fiscal climate risks through access to climate-sensitive in advance disaster risk finance facilities, or contingent disaster financing, as well as catastrophe bonds for infrastructure investment at the national and regional level. ADB will support the integration of climate aspects in an existing regional program, Achieving Climate Change Objectives through Public Financial Management Reforms. Project example is Supporting Regional Actions to Address Climate Change under the Central Asia Regional Economic Cooperation Program. For further reading, see R. Jonassen et al. 2023. Carbon Pricing and Fossil Fuel Rationalization Tool Kit. ADB.

33 This will be done through the Coalition of Finance Ministers for Climate Action.

48. **Support the mobilization of private climate investment.** ADB will support private sector alignment of financial flows with the transition to net zero and climate-resilient development. This will be done through support for local financial institutions' efforts to decarbonize and de-risk portfolios and support for their clients' efforts to transition and build resilience. ADB will support the enhancement of a conducive regulatory framework and business environment including for PPPs.[34] The bank will promote the deployment of effective and efficient fiscal stimuli promoting private investment in energy efficiency measures and green investments. It will focus on private sector operators for public infrastructure, agribusiness, manufacturing, other industry, mining, and real estate. It will continue promoting green and sustainable finance development in DMCs through the development of an environmental, social, and governance framework; a national green and sustainable taxonomy; and investment instruments such as public or private green and blue bonds. This will be accompanied by capacity building and training for private businesses large and small, as well as for state-owned enterprises, on international nonfinancial reporting and disclosure requirements. For project examples, see Building Institutional Capacity: Delivering Climate Solutions under Operational Priority 3 of Strategy 2030 and Promoting Sustainable Public–Private Partnerships Program in Pakistan.

49. **Support research and development, the establishment of innovation hubs, and the formation of partnerships between universities and research institutions.** ADB will promote innovation and technology transfer within the region. It will explore ways to introduce and/or disseminate the adoption of cutting-edge technologies, such as blockchain for carbon trading, or deploying artificial intelligence for climate data analysis and policymaking.

50. **Develop climate-smart public infrastructure and design standards for the energy, transport, and water sectors.** ADB will continue to help governments develop energy-efficient, low-emission industry and construction standards at the national and regional level. This will help meet climate requirements for infrastructure, thereby enhancing infrastructure resilience in the face of climate and natural hazards. Public services such as water supply, electricity generation, and public transport will be decarbonized.

51. Table 8 provides entry points through which ADB can help DMCs align their climate policies with the Paris Agreement. Table 9 provides entry points for climate-related financial systems.

[34] Encompassing PPP laws and regulatory frameworks, the preparation of public tender processes, technical assistance for bidders to develop bankable investment projects, and power purchase agreements.

Table 8: Climate-Related Interventions Supporting Governance, Institutions, and Public Sector Policy Reforms

Intervention	Entry Point	Armenia	Azerbaijan	Georgia	Kazakhstan	Kyrgyz Republic	Pakistan	Tajikistan	Turkmenistan	Uzbekistan	Short Term 1–2 Years / Long Term to 2030
Develop climate policy frameworks and mainstream climate actions in the PSMG operational pipeline.											
Support NDC update for 2025.	Upstream	H	H		H	H	H	H	H	H	Short term
Support general capacity building toward climate change implications on sectors and climate policies among government counterparts in DMCs.	Upstream	H	H	x	H	H	H	H	H	H	Short term
Support international carbon markets under Paris Agreement Article 6.	Upstream	x		x		x	x	x		x	Short term
Develop carbon pricing policies, tool kits, emissions trading systems, and carbon taxation.	Upstream	x	x	x	x	x	x		x	x	Long term
Support the development of the Enhanced Transparency Framework under Paris Agreement Article 13.	Upstream	x	x	x	x	x	x	x	x	x	Long term
Develop inclusive low-carbon development strategies such as for long-term low-emission development, a green economy transition, and a just transition.	Upstream		x	x	x	x		x	x	x	Long term
Support interagency coordination through existing climate councils and capacity building and institutional networks toward national climate policies.	Upstream	x	x	x	x	x	x	x	x	H	Short term
Enhance national climate change mitigation NDCs, NAPs, and road maps.	Midstream	H	H	H	H	H	H	H	H	H	Long term
Support DMCs' formulation of fiscal and public financial management policy actions.											
Support sovereign debt management for climate finance through, for example, debt-for-nature swaps (CCAP actions B.2.3 and E.1.2).	Upstream					x	x	x		x	Short term
Support the implementation of climate public investment management assessment (CCAP action E.1.2).	Upstream	x		x		x	x	x		x	Short term
Promote the Coalition of Finance Ministers for regional capacity building for climate-related public management policies and systems (CCAP action E.1.2).	Upstream				x	x				x	Long term

continued on next page

Table 8 continued

Intervention	Entry Point	Armenia	Azerbaijan	Georgia	Kazakhstan	Kyrgyz Republic	Pakistan	Tajikistan	Turkmenistan	Uzbekistan	Short Term 1–2 Years / Long Term to 2030
Promote the greening of public financial management and climate integration (CCAP action E.1.2).	Upstream	x	x	x	x	x	x	x	x	x	Long term
Support analysis and reporting of fiscal risks through climate change (CCAP action E.1.2).	Midstream			H		x		x		x	Long term
Promote the legal, regulatory, and fiscal climate framework in selected economic and social sectors.											
Develop layered, in advance, disaster risk finance through contingent disaster financing and catastrophe bonds for infrastructure investment.	Upstream	x				x	x	x		x	Long term
Promote gender-sensitive policies for women's employment in the energy sector.	Upstream					x		x		x	Long term
Promote transboundary cooperation on disaster and climate risk management.	Upstream		x	x	x	x		x	x	x	Long term
Support the development of a green economy and circular economy strategies.	Upstream	x				x		x		x	Long term
Foster innovation and technology transfer through research and development, new innovation hubs, partnerships between universities and research institutions, and the introduction of new technologies (CCAP action E.1.3).	Midstream				x					x	Long term
Develop climate-smart public infrastructure and design standards for the energy, transport, and water sectors.	Midstream	x			x	x		x	x	x	Long term
Promote reform in state-owned enterprises and requirements for private sector environmental, social, and governance reporting and disclosure.	Downstream		x	x	x	x			x	x	Long term

CCAP = *Climate Change Action Plan, 2023–2030*; DMC = developing member country; NAP = national adaptation plan; NDC = nationally determined contribution; PSMG = public sector management and governance.

Note: High priorities are identified with an H, and medium priorities with an x. Blank indicates that the action was not identified as a priority through analysis and engagement with the countries but could become a priority in coming years.

Source: Asian Development Bank.

Table 9: Climate-Related Financial System Interventions

Intervention	Entry Point	Armenia	Azerbaijan	Georgia	Kazakhstan	Kyrgyz Republic	Pakistan	Tajikistan	Turkmenistan	Uzbekistan	Short Term 1–2 Years / Long Term to 2030
Support initiatives on green, sustainable transition finance.											
Promote the development of sustainable green finance frameworks (CCAP action D.3.1).	Upstream	H	H		H	x		x		x	Short term
Support capacity building on a national green taxonomy, IFRS S1/S2, and ESG standards (CCAP action D.3.1).	Midstream	x	x	H	H	x	x	x	x	x	Short term
Support regulatory frameworks for ESG-related development of capital markets and debt financing through green investment funds, thematic bonds, and guarantees to attract domestic investors (CCAP actions D.3.2 and D.3.3).	Midstream	x	x	x	H			x		x	Short term
Promote feasibility assessments and the enhancement of regulatory frameworks for public and private green investment to raise private capital, such as green and sustainability bonds (CCAP actions D.3.2 and D.3.3).	Midstream	x		x	x	x		x		x	Short term
Promote ESG finance and ESG reporting and disclosure requirements for corporations, financial institutions, and state-owned enterprises, such as GHG footprint analyses (CCAP action D.2.2).	Downstream	x	x	x	H	x	x	x	x	x	Short term
Support selected private companies' access to green capital market opportunities such as green investment, green borrowing, and green bonds (CCAP actions D.3.2 and D.3.3).	Downstream	x		H	x	x		x		x	Short term
Support selected financial institutions' greening of their lending portfolios with reduced interest rates and cashbacks (CCAP actions D.3.2, D.3.3, and G.1.7).	Downstream	x	x	x	x	x	x	x		x	Short term
Support life-cycle-based green housing finance (CCAP actions D.3.2, D.3.3, and G.1.7).	Downstream	x			x	x					Long term
Prepare frameworks for disasters, pandemic risk financing, and feasibility studies on disaster insurance products.											
Develop climate and disaster insurance solutions in partnership with the private sector (CCAP action G.1.7).	Downstream	x			H		x		x		Long term

CCAP = *Climate Change Action Plan, 2023–2030*; ESG = environmental, social, and governance; IFRS = International Financial Reporting Standards.

Note: High priorities are identified with an H, and medium priorities with an x. Blank indicates that the action was not identified as a priority through analysis and engagement with the countries but could become a priority in coming years.

Source: Asian Development Bank.

52. **Challenges and limitations.** A major challenge is to align national climate policies with sector-specific and gender-sensitive action plans, which requires significant coordination among various government institutions and stakeholders. Additionally, the decentralization of national climate policies into subnational plans can be particularly challenging in countries with highly decentralized governance structures, such as Pakistan, where local governments may lack the capacity or resources to effectively implement these policies. Another limitation is the integration of climate policies into larger fiscal frameworks, which involves complex processes like developing carbon pricing policies and managing public financial and investment systems to support low-emission initiatives. Further, mobilizing private sector investment for climate action requires overcoming barriers related to the general investment and business climate in DMCs, as well as prevailing financial risks and market incentives, which can be difficult in some economies.

Green, Resilient, Integrated, and Diversified Economic Growth

53. **Scale up investment and mobilize new types of transformation-enabling energy infrastructure.** ADB will continue to help DMCs increase the share of renewable energy sources in the national energy mix. The bank will develop and invest in renewable energy systems and associated infrastructure such as smart and resilient power grids; battery energy storage; investment in mini and micro hydropower stations; wind power energy infrastructure; piloting green hydrogen generation, transport, and storage; floating solar installations; and investment in digital infrastructure for enhanced management of energy systems. For project examples, see Bukhara Solar and Battery Energy Storage Project in Uzbekistan and, in Kazakhstan, Baikonyr Solar Project, Total Eren Access M-KAT Solar Power Project, and Samruk-Energy Green Transformation.

54. **Mobilize and scale up investment to design, replace, and retrofit climate-vulnerable infrastructure.** Climate change affects public infrastructure not only through acute physical hazards—or disasters triggered by natural hazards such as river floods, strong winds, and landslides—but also through slow-onset changes, such as thawing permafrost soil or sinking groundwater tables. ADB will help partner with countries to conduct sector or geographic climate risk assessments to inform project design concepts and upgrade or rehabilitate climate-vulnerable infrastructure such as road and rail networks, power transmission and distribution systems, and water supply infrastructure to ensure that critical infrastructure is resilient under current and future climate risks.[35] For a project example, see the Emergency Flood Assistance Project in Pakistan.

55. **Pursue deep decarbonization of large emitters in key industries.** ADB will help partner countries shift away from coal-fired power and heating plants to ensure a gender-sensitive, just, and fair transition for people and communities. Large state-owned enterprises in, for example, cement, aluminum, steel, and fertilizer production will require support as they replace high emission technologies with energy-efficient and low-carbon solutions, such as replacing coal-fired power plants in Kazakhstan by deploying ADB's Energy Transition Mechanism, and to respond to the European Union's CBAM.[36] Improved corporate GHG balance will be integrated into corporate environmental, social, and governance reporting and disclosure, as well as the national GHG inventory. Similarly,

[35] P. Watkiss et al. 2020. Principles of Climate Risk Management for Climate Proofing Projects. *ADB Sustainable Development Working Paper Series*. No. 69. ADB.

[36] One of this mechanism's core pillars is a just transition. Activities under this pillar aim to assess and mitigate direct, indirect, and induced impacts at the asset level. These actions are typically coordinated with the national just transition approach.

ADB will help the private sector finance upgrades or retrofits for energy efficiency and fuel-switching solutions in manufacturing and construction, and leverage green-equipment facilities to support the import and distribution of low-carbon technologies. For project examples, see the Energy Transition Mechanism technical assistance project Opportunities to Accelerate Coal to Clean Power Transition and the ALES Energy Transition and Modernization Project, both in Kazakhstan.

56. **Crowd in and mobilize private sector investment for public services and utilities, including renewable energy, water, and transport infrastructure.** ADB helps DMCs enhance their regulatory frameworks and business conditions to attract private investment for partnerships with the public sector. Investors require viable and economically attractive power-purchase agreements, feed-in tariffs, or off-taker arrangements. Aside from enhancing the regulatory framework, this will encompass supporting the development of PPPs, as well as measures to de-risk private investment through, for example, sovereign first-loss guarantees and blended finance instruments. This includes working with other development organizations, donors, and funds to catalyze private sector investment. On the investor side, private partners will be helped to participate successfully in public tender processes and introduce digital infrastructure for the management of public services and utilities. For project examples, see Solar Public–Private Partnership Investment Program in Uzbekistan and Advancement of Private Sector Participation and Development of the Roadmap for Public–Private Partnerships in Turkmenistan.

57. **Systemically decarbonize the transport sector through e-mobility and the development of multimodal transport systems on economic corridors.** ADB aims to enhance climate-resilient, low-carbon transport by shifting from fossil fuels to electric systems and integrating efficient multimodal transport. Support will focus on low-emission vehicles, resilient infrastructure, hydrogen and other low-emission power trains and rail systems, skills development for all genders, policy and financial support, and capacity building for full-scale decarbonization. ADB will back low-carbon trade and supply chains for electric vehicles, positioning Uzbekistan as a regional e-vehicle producer. Efforts include updating education for skills in manufacturing and maintaining electric vehicles and promoting female employment. ADB will assist in transitioning transport systems to renewable energy, upgrading electric infrastructure, and increasing renewable energy use in transport across several countries.[37]

58. **Greening the financial system.** ADB recognizes the importance of the financial system in transmitting climate policies to economic entities and thus mobilizing private investment for low-carbon and resilient economic development. ADB will help DMCs develop sustainable and green finance frameworks featuring green taxonomy and environmental, social, and governance standards, and enhance the regulatory framework for green investment funds, bonds, and guarantees attracting investment from domestic investors, such as insurance and pension funds. Further, ADB will support financial institutions and businesses large and small on International Financial Reporting Standards S1/S2 reporting and disclosure standards, thus encouraging further private climate investment. ADB can support selected financial institutions in the greening of their lending portfolios and financial intermediation to deliver positive social and environmental impacts. ADB will align its support to regional policy frameworks through the gradual adoption of legislation and regulatory elements subject to the EU's so-called *acquis communautaire* (body of law) in West Asia under specific trade agreements, such as the Comprehensive and Enhanced Partnership in Armenia and the Deep and

[37] For further reading, see ADB. 2016. *Unlocking the Potential of Railways. A Railway Strategy for CAREC, 2017-2030*; and ADB. 2020. *CAREC Transport Strategy 2030*.

Comprehensive Free Trade Area in Georgia, and facilitate accession candidate status for Georgia. For project examples, see Georgian Green Bond 2 Project and Georgia Capital Sustainability-Linked Bond Project, both on debt security in Georgia, and Enabling Green Recovery in Central and West Asia through a Sustainable Financing Program in Armenia, Kazakhstan, Pakistan, and Uzbekistan.

59. **Develop sovereign or private climate and disaster risk transfer instruments.** ADB will help DMCs create climate and disaster insurance for climate and natural hazards. This includes developing a supportive framework with regulatory authorities, involving reinsurance companies, and creating indemnity or index-based risk transfer with the private sector. Regionally, through CAREC, ADB supports an innovative disaster risk transfer facility involving several countries, focusing on capacity building and feasibility analyses. The mechanism includes issuing catastrophe bonds, integrating them with public fiscal management, and exploring nonsovereign risk management for the private sector. ADB will also pilot a disaster relief catastrophe bond in selected CAREC countries. For project examples, see Strengthening the Enabling Environment for Disaster Risk Financing and Developing a Disaster Risk Transfer Facility in the Central Asia Regional Economic Cooperation Region.

60. Tables 10 and 11 detail entry points for increasing the share of renewable energy, enhancing the climate resilience of infrastructure, and supporting the decarbonization of major industries.

Table 10: Climate-Related Energy Interventions

Intervention	Entry Point	Armenia	Azerbaijan	Georgia	Kazakhstan	Kyrgyz Republic	Pakistan	Tajikistan	Turkmenistan	Uzbekistan	Short Term 1–2 Years / Long Term to 2030
Support private sector involvement to expand low-carbon energy.											
Support alignment of energy tariffs for renewable energy competitiveness.	Upstream	x	H		x	x	H	H		x	Long term
Support regulatory frameworks for regional power trade agreements, low-carbon power purchase agreements, feed-in tariffs, and off-taker arrangements.	Midstream		H	x	x	x		H	x	x	Long term
Support the development of de-risking mechanisms for renewable energy investors, such as sovereign first-loss and off-taker guarantees.	Midstream				x	x		x	x		Long term
Facilitate tender processes for renewable energy infrastructure construction and operation concessions.	Midstream	H			x	x		x	x		Long term
Promote public–private partnership frameworks for small and medium-sized solar, wind, hydropower production, distribution, and storage	Midstream	x			H	x		x	H	H	Long term

continued on next page

Table 10 continued

Intervention	Entry Point	Armenia	Azerbaijan	Georgia	Kazakhstan	Kyrgyz Republic	Pakistan	Tajikistan	Turkmenistan	Uzbekistan	Short Term 1–2 Years / Long Term to 2030
Prepare renewable energy investors for bidding processes	Downstream					H		H		x	Long term
Enhance climate resilience across the energy sector.											
Analyze climate hazards to inform the energy sector.	Upstream	x	x	x	x	x	x	x	x	x	Short term
Support institutional development and good governance and assist in creating policy frameworks to manage the energy transition.	Midstream		x		H	x		x	x	H	Long term
Support the development of sector adaptation plans.	Midstream	x	x	x	x	x	x	x	x	x	Long term
Develop climate and disaster insurance for the energy sector in partnership with the private sector.	Downstream	x	x	x			H				Long term
Upgrade power plants and networks against climate hazards to improve, for example, hydropower resilience when flooded and transmission resilience in the face of storms and landslides.	Downstream	x	x	x	x	x	x	x	x	x	Long term
Facilitate resilience planning and response for operations.	Downstream	x	x	x	x	x	x	x	x	x	Long term
Invest in new climate- and disaster-resilient infrastructure.	Downstream	x	x	x	x		H	x		x	Long term
Support decarbonization in the energy sector.											
Develop policies and regulations to support low-carbon technologies and introduce market-based instruments including carbon pricing.	Upstream	x	x	x	x	x	x		x	x	Long term
Develop green hydrogen road maps (CCAP action G.1.4).	Upstream			x	H						Short term
Support decarbonization by closing fossil fuel thermal power plants and transitioning to non-fossil technologies under the Energy Transition Mechanism (CCAP action G.1.4).	Upstream	x	x	x	H	x	H	x	x	x	Long term
Promote regional energy cooperation to strengthen energy security and increase cross-border access to cleaner energy sources.	Upstream	H	x	H	H	H	x	H	x	H	Long term
Set operational standards to reduce methane emissions from natural gas distribution networks.	Midstream		x						x		Long term

continued on next page

Table 10 continued

Intervention	Entry Point	Armenia	Azerbaijan	Georgia	Kazakhstan	Kyrgyz Republic	Pakistan	Tajikistan	Turkmenistan	Uzbekistan	Short Term 1–2 Years / Long Term to 2030
Analyze the feasibility of innovative renewable technologies such as wind parks, heat pumps, and green hydrogen (CCAP action G.1.4).	Midstream	x		x	x						Short term
Invest in innovative renewable energy infrastructure such as floating solar and energy storage in batteries (CCAP action G.1.4).	Downstream			x	x					x	Long-term
Invest in mini and micro hydropower plants and renewable energy wind power infrastructure.	Downstream	x	x	x	H	H	x	H	x	x	Long-term
Stabilize power grid supply and demand (CCAP action G.1.4).	Downstream	H				x	x	x		x	Long term
Pilot invest in green hydrogen generation and storage (CCAP action G.1.4).	Downstream		x	x	x					x	Long term
Transition high-emission corporate and SOE business processes to low-carbon technologies and support SOEs' ESG reporting capacity.	Downstream				x	x	H	x	x	x	Long term
Deploy RE for desalination plants (CCAP action G.1.4).	Downstream				H				x		Long term
Support energy efficiency and energy management.											
Deploy renewable energy for desalination plants (CCAP action G.1.4).	Midstream	x			x	x		H			Long term
Support directives and regulations on energy efficiency in maritime transport.	Midstream			H							Long term
Enhance efficiency in existing power plants and networks, including hydropower plants, to reduce losses.	Downstream		x		x	x		x	x	x	Long term
Support efficiency and resource recovery in industrial processes.	Downstream		x	x	x	x	x		x	x	Long term

CCAP = *Climate Change Action Plan, 2023–2030*; ESG = environmental, social, and governance; SOE = state-owned enterprise.

Note: High priorities are identified with an H, and medium priorities with an x. Blank indicates that the action was not identified as a priority through analysis and engagement with the countries but could become a priority in coming years.

Source: Asian Development Bank.

Table 11: Climate-Related Transport Interventions

Intervention	Entry Point	Armenia	Azerbaijan	Georgia	Kazakhstan	Kyrgyz Republic	Pakistan	Tajikistan	Turkmenistan	Uzbekistan	Short Term 1–2 Years / Long Term to 2030
Involve the private sector in low-carbon and resilient transport systems.											
Utilize financing modalities to promote private sector investment in low-carbon and resilient public transport.	Midstream	x		x	x	x		x	x	x	Long term
Promote private sector investment, including through public–private partnerships, in low-carbon and resilient transport infrastructure construction, maintenance, and operation.	Midstream	x		x	x	x		x		x	Long term
Promote private sector support, including through public–private partnerships, for enhanced supply chains for e-vehicle production and regional distribution.	Midstream	x	x	x	x	x	x	x	x	x	Long term
Develop green skills and TVET curricula for low-emission private and public transport technologies.	Midstream	x		x	x	x		x		x	Long term
Decarbonize accessible and integrated road, rail, air, and water transport systems.											
Support feasibility studies for road, rail, aviation, and maritime transport decarbonization (CCAP actions G.1.4 and G.1.5).	Upstream	x	x	x	x	x	x	x	x	x	Long term
Support a mode shift to public transport, particularly to efficient urban transport.	Midstream	x	x	x	x	x	H	x	x	x	Long term
Establish minimum fuel standards, pollution controls, and enhanced producer responsibilities for motor vehicles, aircraft, and watercraft.	Midstream		x							x	Long term
Provide policy advice on tax and fiscal incentives for e-vehicle promotion and promote national policies on circular economies for vehicle battery recycling and disposal.	Midstream	x		x	x	x	x	x		x	Long term
Develop digital ticketing solutions to encourage the use of public transport.	Midstream	x		x		x		x		x	Long term
Revise municipal revenues, fares, and transport prices for sustainable public transport reform.	Midstream	x		x	x	x		x		x	Long term
Promote efficient traffic control systems to reduce traffic congestion and associated emissions.	Midstream		x	x	H			x	H		Long term
Promote low-carbon and efficient freight, logistics, and supply chain operations.	Midstream	x	x	x	x					x	Long term

continued on next page

Table 11 continued

Intervention	Entry Point	Armenia	Azerbaijan	Georgia	Kazakhstan	Kyrgyz Republic	Pakistan	Tajikistan	Turkmenistan	Uzbekistan	Short Term 1–2 Years / Long Term to 2030
Promote CAREC CCAP measures for decarbonization.	Midstream		x	x	x	x	x	x	x	x	Long term
Adjust energy mix for e-transport systems (CCAP action G.1.4).	Midstream	x		x	x	x		x		x	Long term
Invest in delivering e-vehicle charging and hydrogen vehicle refueling infrastructure.	Downstream			x	x	x		x		x	Long term
Substitute diesel-driven vehicles through overhead powerline, battery, or green hydrogen vehicles (CCAP action G.1.4).	Downstream	H	x	H	H	x	H	x	x	H	Long term
Promote and invest in infrastructure for safe cycling and walking, particularly in urban areas.	Downstream			x						x	Long term
Promote climate- and hazard-resilient transport infrastructure.											
Revise construction guidelines and standards for climate- and disaster-resilient transport infrastructure.	Upstream				x	x		x	x	x	Short term
Develop and enhance resilient regional cross-border transport corridors, particularly missing links.	Midstream	x	x	x	x	x		x	x	x	Long term
Promote digital solutions for efficient and resilient asset management and maintenance.	Midstream	x	x	x	x						Long term
Support capacity development for resilient transport operations, in particular for public transport providers.	Midstream					x		x		x	Short term
Invest in climate- and disaster-resilient transport infrastructure, including resilient public transport and the enhancement of vulnerable infrastructure.	Downstream	x	x	x	x	x	H	x	x	x	Long term
Promote nature-based solutions for roadsides to improve air quality and support sustainable road drainage.	Downstream					x		x		x	Long term
Invest in real-time monitoring of transport infrastructure demand and performance, integrated with climate hazard warning systems.	Downstream				x	x		x		x	Long term
Invest in transport corridors and services that provide access to critical services, such as disaster evacuation and response and access to health services.	Downstream				x		x			x	Long term

CAREC = Central Asia Regional Economic Cooperation; CCAP = *Climate Change Action Plan, 2023–2030*; TVET = technical and vocational education and training.

Note: High priorities are identified with an H, and medium priorities with an x. Blank indicates that the action was not identified as a priority through analysis and engagement with the countries but could become a priority in coming years.

Source: Asian Development Bank.

61. **Challenges and assumptions.** A significant challenge is the need for substantial financial resources to support the transition to renewable energy and the development of resilient infrastructure. Mobilizing these resources requires enhancing regulatory frameworks and attracting private sector investment, which can be complex and long term. Additionally, there is a need to ensure that the transition is socially equitable, particularly in the phasing out of coal-fired power plants, which involves addressing potential job losses and economic impacts on communities and cities in, for example, Kazakhstan and Uzbekistan, which rely heavily on fossil fuels or high-carbon industries. Further, the integration of new technologies, such as e-mobility and green hydrogen, requires significant investment in infrastructure and skills development. Lastly, diverse climate risks across regions necessitate tailored approaches, complicating the implementation of support to DMCs' financial systems as they transition toward green and low-carbon economies.

Resilient and Inclusive Societies

62. **Decarbonize public transport in metropolitan areas and urban centers.** ADB will further enhance cooperation with global development partners, notably the GCF and Global Environment Facility, to support green urban planning and the decarbonization of urban public transport in CWA. Aside from investments such as roads, depots, and charging and grid-strengthening infrastructure, ADB will support gender-sensitive vocational training and education for the construction, repair, and maintenance of infrastructure; the development of e-ticketing; and the revision of fares and transport fees for sustainable municipal transport reform. Further, municipalities shall be advised to mobilize and integrate private sector investment in green public transport. Digital systems for the management of public transport, such as efficient parking and traffic control systems, will support the transition from private to public, gender-equitable commuting preferences in climate-smart cities. For a project example, see Urban Transport Electrification Project in the Kyrgyz Republic.

63. **Reform urban policies and governance for greener and more livable cities.** ADB will help DMCs develop low-carbon, climate-resilient urban areas by collaborating with sector ministries on sustainable investments, including nature-based solutions and disaster finance integration. Training officials on climate financing is vital for support. Urban planning will be inclusive and focused on transit-oriented development. In Armenia, Georgia, and the Kyrgyz Republic, ADB partners with the public and private sector to improve urban water supply, sanitation services, and solid waste management, as in Tajikistan. Participatory planning and digital technologies will enhance community involvement. ADB aids CWA in creating green cities through regulatory frameworks and financial innovation. Incentivizing the private sector for energy efficiency and digitizing services are key strategies. For project examples, see Growth, Resilience, Energy Efficiency, and Nature-Based (GREEN) Solutions for Livable Cities; Leveraging the Private Sector to Deliver Inclusive, Affordable, and Climate-Resilient Housing in Central and West Asia; Yerevan Urban Development Investment Project in Armenia; Sustainable Water Supply and Sanitation Sector Development Program in Georgia; Promoting Low-Carbon Development in Central Asia Regional Economic Cooperation Program Cities; and Supporting Development of Innovative Green Housing Finance in Kazakhstan.

64. **Promote financial innovations and mobilize more climate finance for urban development.** To attract more climate finance for urban development, it is essential to enhance the enabling environment for investment in urban infrastructure. Improving the financial sustainability and creditworthiness of municipalities can make them more appealing to investors. Key tools are innovative financing methods such as climate bonds, disaster risk bonds, carbon credits, and blended finance,

which expand access to finance, mitigate risks, and leverage private capital. Additionally, promoting municipal green investment funds, such as in Tashkent, Uzbekistan, and supporting municipal green bonds can direct funding to local climate projects, facilitating investment in renewable energy and health services. For project examples, see Support to Balanced and Sustainable Urban Operations in Central and West Asia and Promoting Low-Carbon Development in Central Asia Regional Economic Cooperation Program Cities for Astana, Kazakhstan.

65. **Promote greener and climate-resilient public services and social infrastructure.** ADB will help CWA countries attract private investment in renewable energy. It will support low-carbon district heating and climate-resilient infrastructure in Azerbaijan, Kazakhstan, the Kyrgyz Republic, and Uzbekistan. In Armenia, Kazakhstan, and Uzbekistan, ADB invests in energy-efficient education and health facilities. The Climate and Health Initiative, established at the 2023 Conference of the Parties of the United Nations Framework Convention on Climate Change (COP28), aims for sustainable, low-carbon health systems, with a focus on gender impacts. In Tajikistan, Kazakhstan, and Uzbekistan, ADB supports low-emission waste management and wastewater treatment facilities. For project examples, see Improve Wastewater Management in Uzbekistan and Supporting Renewable Technology-Inclusive Heat Supply Legislation in Kazakhstan.

66. **Protect people and the environment by enhancing waste management and the circular economy.** ADB actively supports partner countries in CWA to enhance waste management and adopt circular economy practices to safeguard people and the environment. ADB's efforts include technical assistance for integrated solid waste management, notably in Azerbaijan, promoting circular economy principles to minimize environmental impacts and maximize resource efficiency. For project examples, see Supporting the Improvement of Management of Hazardous Waste in Armenia and Sustainable Solid-Waste Management Project in Uzbekistan.

67. **Support shock-responsive and adaptive social protection systems for climate resilience and a just transition.** Adaptive social assistance programs can mitigate climate risks, ensuring robust social protection systems. In the Kyrgyz Republic and Tajikistan, ADB invests in social protection and disaster risk management to shield vulnerable populations from climate change. Developing shock-responsive and adaptive systems is crucial for a just transition. Prioritized groups include children, older people, people with disabilities, and women, as they are more affected by climate change. For project examples, see Building Resilience with Active Countercyclical Expenditures Program in Pakistan, Tajikistan, and the Kyrgyz Republic.

68. **Make disaster risk finance available for shock-responsive social protection systems.** Disaster risk finance for social protection is vital for enhancing community resilience in emergencies. ADB's contingent disaster financing facility aids countries with fiscal constraints, notably Armenia, the Kyrgyz Republic, Pakistan, and Tajikistan, by providing structured financial responses to disasters. This ensures that funds are available when needed most. Integrating regional disaster risk finance mechanisms, such as catastrophe bonds or ADB's contingent disaster financing facility, offers a proactive approach to securing financial resources, stabilizing economies, and aiding swift recovery. For a project example, see Developing a Disaster Risk Transfer Facility in the Central Asia Regional Economic Cooperation Region.

69. **Develop and deliver climate education programs.** In Armenia, Georgia, and Tajikistan, ADB promotes climate education in science, technology, engineering, and mathematics curricula, especially to engage young women. Efforts begin with green labor market assessments and laying the foundations for a just transition. Climate change content is integrated into technical and higher education, emphasizing greening education. This involves incorporating climate education into all educational levels, including teaching materials. Cross-sector solutions are crucial for green workforce development and community resilience, achieved through green investment, social protection, and climate adaptation projects. Supporting higher education in research and innovation fosters sustainable technology development and adoption. For project examples, see Science, Technology, Engineering and Mathematics in Secondary Education Project in Uzbekistan and Improving the Science, Technology, Engineering, and Mathematics Secondary Education Project in Tajikistan.

70. Tables 12 provides entry points for creating resilient and inclusive urban environments. Tables 13–16 focus on enhancing social infrastructure and implementing comprehensive climate education and health programs.

Table 12: Climate-Related Water and Urban Development Interventions

Intervention	Entry Point	Armenia	Azerbaijan	Georgia	Kazakhstan	Kyrgyz Republic	Pakistan	Tajikistan	Turkmenistan	Uzbekistan	Short Term 1–2 years / Long Term to 2030
Improve water and energy efficiency in residential, public, commercial buildings.											
Support regulation such as construction codes and legislation on energy and water efficiency in residential, public, and commercial buildings (CCAP action G.1.3).	Midstream	x	x	x	H	H	x	H	H	H	Long term
Support energy service companies and the private sector for energy efficiency retrofitting investment in public and private buildings and energy and water efficiency in residential investments, e.g., through mortgages.	Midstream	x	x	x	x	x	x	x	x	x	Long term
Support climate change adaptation and disaster risk reduction for urban development.											
Strengthen risk-informed land-use planning (CCAP action G.1.2).	Upstream	x	x	x	x	x	H	x	x	x	Short term
Deploy a contingent disaster financing facility for municipal and national disaster risk management of water and urban infrastructure.	Upstream	x				x	x				Short term
Support climate- and disaster-resilient urban infrastructure planning, risk management, and early warning systems.	Midstream	x	x	x	x	x	H	x	x	x	Long term

continued on next page

Table 12 continued

Intervention	Entry Point	Armenia	Azerbaijan	Georgia	Kazakhstan	Kyrgyz Republic	Pakistan	Tajikistan	Turkmenistan	Uzbekistan	Short Term 1–2 years / Long Term to 2030
Tackle future water scarcity by improving water efficiency, reuse, recycling, and demand management.	Midstream	H	H	x	H	H	H	x	H	H	Long term
Promote green investment funds, green bonds, and special purpose vehicles for urban investments at the municipal level.	Midstream				x	x		x	x	x	Long term
Invest in climate- and disaster-resilient public urban infrastructure, including seismic-safe schools, transport infrastructure, and health and educational facilities, as well as urban flood management and drainage systems.	Downstream	x		x		x	x	x		x	Long term
Support low-emission public services and utilities.											
Support adjustment of tariffs and subsidies to enable private investment in renewable energy.	Upstream						x			x	Long term
Support district heating planning in partnership with the private sector.	Midstream		x		x					x	Long term
Support the digitalization of governmental services and e-government (CCAP action G.1.2).	Midstream	x			x	x		x		x	Short term
Support urban e-mobility planning for public and private transport.	Midstream	H	H	H	H	H	x	H	x	H	Short term
Support public–private investment in solid waste and wastewater treatment facilities to reduce methane emissions.	Downstream	x	x	x	x	x	H	x	x	x	Long term

CCAP = *Climate Change Action Plan, 2023–2030.*

Note: High priorities are identified with an H, and medium priorities with an x. Blank indicates that the action was not identified as a priority through analysis and engagement with the countries but could become a priority in coming years.

Source: Asian Development Bank.

Table 13: Climate-Related Disaster Risk Management Interventions

Intervention	Entry Point	Armenia	Azerbaijan	Georgia	Kazakhstan	Kyrgyz Republic	Pakistan	Tajikistan	Turkmenistan	Uzbekistan	Short Term 1–2 years / Long Term to 2030
Assess hazard and vulnerability risks to infrastructure and communities.											
Promote the CAREC Working Group on Climate Change for the development of regional climate change initiatives and knowledge sharing.	Upstream		x	x	x	x	x	x	x	x	Short term
Support national, subnational, and regional CAREC hazard, climate risk, and vulnerability assessments.	Midstream	x	x	x		x	x	x	x		Short term
Facilitate the alignment and amalgamation of social protection and disaster risk management systems.	Midstream	x				x		x			Long term
Support integrated disaster vulnerability indexes for social protection programs.	Midstream	x				x	x	x			Long term
Support hydromet service development and climate-specific disaster risk management.	Midstream	x	x	x	x	x	x	x	x	x	Long term
Strengthen resilience in public infrastructure to enhance community resilience.	Downstream	x	x	x	x	x	x	x	x	x	Long term
Support continuity plans for health facilities and schools to address impacts from extreme events and climate emergencies.	Downstream	x				x		x		x	Long term

CAREC = Central Asia Regional Economic Cooperation.

Note: High priorities are identified with an H, and medium priorities with an x. Blank indicates that the action was not identified as a priority through analysis and engagement with the countries but could become a priority in coming years.

Source: Asian Development Bank.

Table 14: Climate-Related Social Protection Interventions

Intervention	Entry Point	Armenia	Azerbaijan	Georgia	Kazakhstan	Kyrgyz Republic	Pakistan	Tajikistan	Turkmenistan	Uzbekistan	Short Term 1–2 years / Long Term to 2030
Support disaster risk finance for social protection.											
Support sector-specific, gender-sensitive vulnerability assessments of climate and natural hazards, and develop a vulnerability scoring model and index.	Upstream	x		x		x	x	x	x		Short term
Promote the Asian Development Bank contingent disaster financing facility.	Upstream	x				x	x	x			Short term
Facilitate contingency and emergency planning	Upstream	H				H	x	H			Long term
Support the integration of domestic resource mobilization and climate risks into social protection systems, and promote adaptive, anticipative social assistance programs.	Upstream	x				x	x	x	x	x	Long term
Promote a regional disaster risk finance mechanism or catastrophe bond.	Upstream					x	x	x	x	x	Long term
Support social protection policies, programs, and systems for climate resilience and a just transition.											
Support social protection system diagnostics and alignment with development partners for coordination of support.	Upstream									x	Short term
Develop a just transition platform with evidence-generation and dissemination for data-driven policymaking.	Midstream	x				x		x			Long term
Ensure a just transition through reskilling and upskilling efforts, green public works programs, social assistance, and active labor market policies.	Midstream	x				x		x			Long term
Strengthen poverty and social assessments to capture climate risks and vulnerabilities.	Downstream	x	x	x	x	x	x	x	x	x	Short term

Note: High priorities are identified with an H, and medium priorities with an x. Blank indicates that the action was not identified as a priority through analysis and engagement with the countries but could become a priority in coming years.

Source: Asian Development Bank.

Table 15: Climate-Related Education Interventions

Intervention	Entry Point	Armenia	Azerbaijan	Georgia	Kazakhstan	Kyrgyz Republic	Pakistan	Tajikistan	Turkmenistan	Uzbekistan	Short Term 1–2 years / Long Term to 2030
Develop and deliver climate education programs.											
Conduct country-level green labor market and just transition diagnostics.	Upstream	x	x	x	x	x		x		x	Short term
Integrate climate change content into secondary education, technical and vocational education and training, and higher education institution curricula, notably through science, technology, engineering, and mathematics.	Upstream	H		x	x	H	x	x		H	Long term
Incorporate climate education into project design across all levels of education.	Midstream	x		x		x		x		x	Short term
Promote green workforce development and community climate-resilience capacity building through adaptation projects.	Midstream	x	x	x	x					x	Short term
Support higher education institutions in their development of research and start-up capacity for sustainable technologies.	Midstream	x				x				x	Short-term
Strengthen green skills development.	Midstream		x	x	x	x	x	x	x	x	Long term
Develop and deliver resilient and green social sector infrastructure in the education sector.											
Build capacity in social sector executing agencies to plan, operate, and maintain disaster-resilient, energy-efficient social infrastructure.	Midstream	x				x		x		x	Short term
Develop regulatory standards and guidelines for energy efficiency and low-emission equipment for climate-smart infrastructure and public and private buildings.	Midstream	x			x	x				x	Long term
Invest in climate-resilient infrastructure, low-carbon school campuses, and low-emission supplies and equipment procurement for the public sector.	Downstream	x				x	x	x	x	x	Long term
Promote greenhouse gas accounting and disclosure processes, promote green accreditation for schools, and support the certification of energy efficiency achievements.	Downstream		x	x		x			x	x	Long term

Note: High priorities are identified with an H, and medium priorities with an x. Blank indicates that the action was not identified as a priority through analysis and engagement with the countries but could become a priority in the coming years.

Source: Asian Development Bank.

Table 16: Climate-Related Health System Interventions

Intervention	Entry Point	Armenia	Azerbaijan	Georgia	Kazakhstan	Kyrgyz Republic	Pakistan	Tajikistan	Turkmenistan	Uzbekistan	Short Term 1–2 Years / Long Term to 2030
Establish a climate and health initiative to develop sustainable, climate-resilient, low-carbon health systems and service delivery (CCAP action A.2.2).											
Develop standardized country diagnostic, assessment tools, and resources on climate and health.	Upstream	x		x	x	x	x	x	x		Short term
Support the development of national adaptation plans for health.	Upstream					x	x	x			Long term
Strengthen policy frameworks and action plans for health and climate change.	Midstream	x	x	x	x	x	x	x	x	x	Short term
Innovate crowdsourcing for renewable energy investment in health facilities.	Midstream							x			Long term
Decarbonize health supply chains and medical waste systems, and deliver sustainable manufacturing solutions for pharmaceuticals.	Midstream	x					x	x			Long term
Support workforce capacity building on the climate–health nexus.	Midstream	x		x	x	x	x	x	x		Long term
Support behavior change communication on health and climate change, including for vulnerable groups.	Midstream	x		x	x	x	x	x		x	Long term
Strengthen disease and climate-sensitive surveillance systems.	Downstream	x		x		x		x			Long term
Develop and deliver resilient and green social sector infrastructure in the health sector.											
Build capacity in social sector executing agencies to plan, operate, and maintain disaster-resilient, energy-efficient social infrastructure.	Midstream	x				x		x		x	Short term
Develop technical guidance notes on climate-smart infrastructure design standards, energy efficiency, and low-emission equipment for public buildings.	Midstream	x			x	x				x	Long term
Invest in climate-resilient infrastructure and low-emission equipment procurement for health facilities, promote greenhouse gas accounting and disclosure processes, and support the certification and accreditation of energy efficiency achievements for health-care service providers.	Downstream	x	x	x		x	x	x	x	x	Short term

CCAP = *Climate Change Action Plan, 2023–2030.*

Note: High priorities are identified with an H, and medium priorities with an x. Blank indicates that the action was not identified as a priority through analysis and engagement with the countries but could become a priority in coming years.

Source: Asian Development Bank.

71. **Challenges and limitations.** A significant challenge for climate-smart urban development is posed by the need for substantial infrastructure investments such as roads and grid transformation, which require coordination with global development partners and the private sector. This can be hindered by commitment and capacity of stakeholders. Additionally, regulatory reform and governance improvement is needed to create greener, more livable cities, which can be complicated by bureaucratic inertia and resistance to change. Mobilizing climate finance presents a challenge, as it requires innovative financial mechanisms, such as municipal climate funds that mobilize private sector investment, and the strengthening of municipal creditworthiness. Further, the transition to renewable energy demands inclusive adjustment of existing utility costs and subsidies, which can be politically sensitive and economically challenging. A primary challenge to the integration of climate risk data into existing social protection systems is the required comprehensive diagnosis and alignment with development partners. Additionally, ensuring that social protection systems are both shock-responsive and adaptive to the evolving nature of climate risks demands significant coordination across multiple sectors and stakeholders. Financial constraints are a limitation, as most DMCs have few contingency resources available for disaster risk management, necessitating innovative solutions for structured financial responses like catastrophe bonds or ADB's contingent disaster finance facility.

Sustainable Management of Natural Resources and the Environment

72. **Enhance sustainable management of natural resources, landscape management, and forest restoration.** ADB supports Georgia, for example, create regulatory frameworks for landscape management, land tenure, and forest restoration to improve natural resource management and mitigate climate and natural hazards. These frameworks support disaster risk management by promoting hazard-resilient practices. Legal frameworks are vital for the sustainable management of landscapes such as grasslands and forests. A national ecosystem-services market encourages conservation and sustainable land use. Nature-based solutions like tree embankments enhance ecological resilience. Additionally, ADB assists Azerbaijan, Georgia, and Pakistan with a blue-bond investment framework for marine ecosystem resilience and supports the water pillar under a CAREC program for improved transboundary water management. For project examples, see National Disaster Risk Management Project in Tajikistan, Developing the Central Asia Regional Economic Cooperation Water Pillar, and Integrating Climate Resilience in the Water Sector.

73. **Promote transboundary and integrated water management.** ADB aids DMCs in CWA by enhancing integrated and transboundary water management for sustainable development and climate resilience. This involves fostering cross-border cooperation, conducting climate risk assessments, and developing cohesive river basin management policies. National efforts include creating guidelines for land and water use, improving subnational river basin management, and supporting demand-driven systems. Financial investments focus on infrastructure like water storage facilities, hydro stations, and desalination plants for irrigation. Additionally, ADB prioritizes water metering, precision irrigation, drainage, salinity reduction, and small-scale water collection to optimize usage and support sustainable agriculture. Box 3 showcases the Glaciers to Farms regional program, which is proposed for GCF cofinancing. For project examples, see the Climate Smart Irrigation Sector Development Program in Georgia, Climate Adaptive Water Resources Management in the Aral Sea Basin Sector Project in Uzbekistan, and Preparing the Climate Adaptive Water Resources Management in the Aral Sea Basin Project. For further reading, see CAREC: Developing the Water Pillar, a 2021 scoping report[38].

[38] CAREC: Developing the Water Pillar—Scoping Report.

Box 3: Proposed ADB Regional Glaciers to Farms Program

The Asian Development Bank's Glaciers to Farms (G2F) regional program aims to promote climate resilience and low-carbon development focused on climate adaptation and sustainable development in Central and West Asia. G2F envisages mobilizing up to $3.5 billion from ADB and other cofinancing partners (e.g., Green Climate Fund) to invest in projects from 2025 to 2031. It will address the region's vulnerabilities to climate change effects, such as glacier melt, water scarcity, and declining agricultural productivity, which threaten socioeconomic stability.

The proposed G2F program has four components: (i) climate resilience through integrated planning and policy reforms for financial acceleration; (ii) glacier investment and technical facility with pillars on climate-resilient landscapes for water resource management, sustainable agriculture value chains and food security, and enhancing adaptation for climate-vulnerable poor populations in mountainous and rural areas; (iii) establishment of green business financing lines for small and medium-sized agriculture enterprises; and (iv) a G2F regional knowledge-sharing and coordination hub.

The G2F program proposes to cover nine developing member countries in Central and West Asia: Armenia, Azerbaijan, Georgia, Kazakhstan, the Kyrgyz Republic, Pakistan, Tajikistan, Turkmenistan, and Uzbekistan. It promotes regional transboundary cooperation and supports the operationalization of the *Climate Action Road Map for Central and West Asia, 2025–2030* and accompanying country climate action plans. Further, it will support countries' implementation of nationally determined contributions and national adaptation plans.

Source: Asian Development Bank.

74. **Develop resilient, sustainable, and gender-sensitive agricultural value chains and livestock management.** ADB helps DMCs build resilient agricultural value chains and livestock management for sustainable growth and climate resilience. Initiatives like regional climate risk transfer facilities help manage risks. Enhancing value chains in CWA through food loss prevention and improved storage reduces GHG emissions and resource consumption. Emphasis on low-carbon methods, such as regenerative agriculture, is crucial, as are formalizing farmer cooperatives, supporting reform of the land code as in Uzbekistan, using digital tools like livestock management systems in Tajikistan, and improving productivity. Investments in climate-smart inputs and finance bolster resilience, supporting environment-friendly practices and infrastructure to mitigate climate risks. For project examples, see Climate-Resilient Agricultural Value Chain Development Project in the Kyrgyz Republic and Enhancing Climate Resilience in the Pyanj River Basin in Tajikistan.

75. Tables 17–19 provide entry points for the sustainable management of landscapes, forests, and water resources, including transboundary water management and climate-smart agricultural practices.

Table 17: Climate-Related Interventions Supporting Sustainable Management of Natural Resources and the Environment

Intervention	Entry Point	Armenia	Azerbaijan	Georgia	Kazakhstan	Kyrgyz Republic	Pakistan	Tajikistan	Turkmenistan	Uzbekistan	Short Term 1–2 Years / Long Term to 2030
Support the development and monitoring of agriculture sector national adaptation plans and low-carbon agricultural development strategies and action plans.	Upstream	H	x				H	x		x	Short and long term
Build research capacity in greenhouse gas monitoring, accounting verification, and the monetization of carbon credits to promote low-carbon agriculture.	Midstream	x					x	x		x	Short term
Develop digital livestock management that supports climate change-sensitive and sustainable use of pasture and other land.	Midstream						x	x		x	Long term
Support the development and distribution of leasing companies toward climate-appropriate products.	Midstream						x	x			Long term
Develop agricultural input supplies and the certification of more resilient seed varieties from nurseries.	Midstream		x				x	x			Long term
Strengthen disaster risk management and investment planning for climate adaptation measures at the national and regional level.	Midstream	x	x	x	x	x	x	x	x	x	Long term
Support environmental and climate information systems for the agriculture sector.	Midstream	x				x	x	x	x		Long term
Support food security and nutrition by building climate resilience, optimizing resource use, preventing food loss, recycling agricultural waste, and boosting storage capacity to sustainably enhance agricultural productivity.	Midstream	x	x	x	x	x	H	x	x	x	Long term
Facilitate farm investment in climate-smart technologies.	Downstream	x			x	x	x	x	x	x	Short term
Invest in tree and plant nurseries.	Downstream			x			x	x			Short term
Invest in early warning system infrastructure against hail, river floods, and landslides.	Downstream	x	x	x		x	x	x	x		Short term
Support reduction in livestock emissions through internal fermentation, manure management, and anaerobic digestion to generate biogas.	Downstream						x	x		x	Long term
Invest in resilient low-emission storage and transport of agricultural produce.	Downstream	x	x				x	x			Long term
Invest in agricultural service facilities and extension services.	Downstream						x	x	x		Long-term
Invest in carbon sequestration through annual crops and the rehabilitation of soil organic matter.	Downstream	x		x		x	x	x			Long term

Note: High priorities are identified with an H, and medium priorities with an x. Blank indicates that the action was not identified as a priority through analysis and engagement with the countries but could become a priority in coming years.

Source: Asian Development Bank.

Table 18: Climate-Related Interventions Supporting Natural Resources, Landscape Management, and Forest Restoration

Intervention	Entry Point	Armenia	Azerbaijan	Georgia	Kazakhstan	Kyrgyz Republic	Pakistan	Tajikistan	Turkmenistan	Uzbekistan	Short Term 1–2 Years / Long Term to 2030
Develop regulatory frameworks for landscape management, land tenure, and forest restoration.	Midstream			x				x			Short term
Support the development of national disaster risk management frameworks.	Midstream			x			x				Short term
Promote hazard-resilient land management and building codes.	Midstream			x	x	x	x	x	x	x	Long term
Support a blue-bond investment framework through the development of an investment project portfolio pipeline.	Midstream			x			x				Long term
Support the development of a national ecosystem services market for agricultural areas, forests, grasslands, and wetlands.	Midstream			x							Long term
Conduct studies to inform sustainable development of the coastal zone and expand sustainable coastal defense using nature-based solutions (CCAP action G.2.2).	Midstream		x	x	x		x		x		Long term
Promote nature-based solutions, such as tree embankments, windbreaks, and rangeland rehabilitation.	Downstream		x	x			x				Short term

CCAP = *Climate Change Action Plan, 2023–2030*; GHG = greenhouse gas.

Note: High priorities are identified with an H, and medium priorities with an x. Blank indicates that the action was not identified as a priority through analysis and engagement with the countries but could become a priority in coming years.

Source: Asian Development Bank.

Table 19: Integrated Transboundary and Domestic Water Management

Intervention	Entry Point	Armenia	Azerbaijan	Georgia	Kazakhstan	Kyrgyz Republic	Pakistan	Tajikistan	Turkmenistan	Uzbekistan	Short Term 1–2 Years/ Long Term to 2030
Promote integrated cross-border and domestic water resource management and climate resilience (CCAP action G.1.6).	Upstream	x	x	x	x	x	x	x	x	x	Short term
Support transboundary and national policy work on river basin management and upstream climate risk assessments at the regional and national level, using CAREC.[a]	Upstream	x	x	x	x	x	x	x	x	x	Long term
Promote integrated water resource and asset management systems for climate-resilient public water infrastructure at the national level.	Midstream	H	H	H		x	H	x	x	x	Long term
Develop guidelines and regulations for land-user and water-user organizations and promote the development of climate-sensitive user fee policies.	Midstream		x	x			x	x	x	x	Long term
Develop subnational river basin management, hydrological modeling, and the management of demand-driven water systems.	Midstream	x	x	x		x	x	x		x	Long term
Promote water efficiency through water metering, reduced conveyance loss, improved farm irrigation efficiency, and thus water conservation.	Downstream				x					x	Long term
Invest in water adaptation upstream in rivers such as water storage, public–private partnerships, early warning systems, and water retention in nature-based solutions.	Downstream				x	x	H	x	x	x	Short term
Invest in water management and desalination of collector-drainage water for agricultural irrigation and salinity reduction.	Downstream		x				x	x	x		Short term
Invest in river basin and landscape management for cross-border cooperation.	Downstream	x	x	x	x	x	x	x	x	x	Short-term
Invest to enhance reservoirs and water storage, embankments, channels, and water collection on smallholder farms, such as in ponds.	Downstream	x		x		x	H	x		x	Long term
Invest in drip irrigation, sprinklers, water metering, and other precision irrigation techniques and infrastructure, as well as capacity building.	Downstream	x	x	x	x	x	x	x	x	x	Long term
Implement soil moisture conservation practices such as mulching, cover cropping, and soil amendment.	Downstream	x	x				x		x	x	Long term

CAREC = Central Asia Regional Economic Cooperation; CCAP = *Climate Change Action Plan, 2023–2030*.

Note: High priorities are identified with an H, and medium priorities with an x. Blank indicates that the action was not identified as a priority through analysis and engagement with the countries but could become a priority in coming years.

[a] The CAREC Working Group on Climate Change will include climate risk assessment aiming to increase resilience in the water sector, with a focus on the water–agriculture–food security nexus. These plans are currently being discussed by the CAREC Working Group on Climate Change for integration into the CAREC CCAP.

Source: Asian Development Bank.

76. **Challenges and assumptions.** A significant challenge is the development of comprehensive regulatory frameworks for landscape management, land tenure, and forest restoration, which require substantial legal and institutional support. This process is complicated by the need for cross-border cooperation and the integration of diverse national policies, particularly in regions like CWA. Another limitation is the financial investment required for infrastructure improvements, such as water storage and irrigation systems, which are crucial for sustainable development and climate resilience. Additionally, the promotion of low-carbon agricultural practices and the establishment of resilient agricultural value chains demand significant changes in current farming practices and market structures, which can be difficult to achieve without comprehensive efforts in capacity building, strong regulatory support, and financial incentives.

Mobilizing Climate Finance

77. ADB is setting ambitious targets for support to CWA DMCs to meet their climate financing needs. It targets mobilizing up to $25 billion in climate finance from 2019 to 2030, including up to $10 billion for adaptation and resilience. This will translate to ADB delivering up to $2 billion in climate finance annually to CWA by 2030. ADB intends that climate financing will expand to key vulnerable and GHG-intensive sectors across DMCs to meaningfully deliver transformative climate actions. ADB will carry out targeted actions to increase support to DMCs, so that greater climate impact is achieved. The actions are outlined in Table 20.

Table 20: Actions to Deliver Greater Climate Impact in Central and West Asia

#	Actions	Measures	Responsible
A.	Ensure that ADB programs and projects respond to DMCs' climate needs and priorities, and embed climate change considerations from the outset.	• Systematically build capacity in sector groups and resident missions so that DMCs' climate needs and priorities are understood, and programs and projects are initiated to respond to these needs, with support from CCSD. • Dedicate internal and external advisory capacity during project conceptualization and planning stages focusing on eligible climate finance measures. • Expand the portfolio toward projects that focus specifically on DMC climate adaptation action (e.g., Type 2a and 2b adaptation), rather than Type 1 climate proofing. • Pursue climate PBLs[a] in those DMCs that require reform to accelerate climate action. It may consider those DMCs with limited sovereign and nonsovereign portfolios at present: Azerbaijan, Kazakhstan, and Turkmenistan. • Continue to strengthen PBLs and increasingly promote sector development programs and results-based loans with coherent alignment toward climate impacts and measurable adaptation, resilience, and mitigation outcomes in Armenia, Georgia, the Kyrgyz Republic, Pakistan, Tajikistan, and Uzbekistan. • Emphasize ADB country allocations and the borrowing limits of DMCs specifically for climate finance, and successfully deploy ADF-14 resources for adaptation and resilience measures (see also points D and E).	• CCSD • CWRD CMTs • Sector groups
B.	Mobilize investment by the private sector to deliver greater climate action.	• Create an enabling environment for private sector investment in climate action and public–private partnership in relevant sectors through policy work such as climate PBLs. • Support private sector bidders in the development of partnerships with the public sector. • Develop competitive and risk-reducing financing instruments for private investors, such as first-loss or sovereign guarantees. • Catalyze private investment in mitigation, such as in charging infrastructure for e-vehicles and adaptation measures.	• PSOD • OPPP • Sector groups • CWRD CMTs

continued on next page

Table 20 continued

#	Actions	Measures	Responsible
C.	Increase collaboration and cofinancing with established and new development partners to deliver greater climate impact.	• Based on DMC climate priorities and ADB's project pipeline, identify joint investment and lending opportunities that align with strategic investment and support priorities of cofinancing partners. • Pursue current multicountry cooperation opportunities at a regional level through, for example, the GCF e-mobility program or the Glaciers to Farms program. • Develop solutions for managing extensive processing timelines through GCF projects.	• CWRD • Sector groups CWRD CMTs • CAREC
D.	Promote grant assistance from the 14th cycle of the ADF to support DMCs with limited financial resources to address climate.	Tajikistan and the Kygryz Republic (ADB lending and grant policy group A: concessional assistance only), as well as Pakistan and Uzbekistan (ADB lending and grant policy group B: ordinary capital resources [OCR] blended for concessionality) are eligible to access ADF-14 grant resources under the thematic windows (i) supporting climate change adaptation and disaster risk reduction, and (ii) fostering regional cooperation and integration, including the provision of regional public goods.[b] For Group A countries, eligible projects can access a maximum two-thirds share of project costs. Considering the need to incentivize projects that aim for transformational outcomes, up to 100% of project costs can be eligible for innovative, transformational, and replicable projects. For Group B countries, a maximum one-quarter share of project costs can be availed by eligible projects. ADF-14 runs from 2025 to 2027.	• Sector groups • CCSD • CWRD CMTs
E.	Increase investment for climate action by accessing existing climate finance facilities.	Promote the utilization of existing climate finance facilities for ADB in CWA such as IF-CAP, Glacier Preservation Agenda, NDC Advance, Nature Solutions Finance Hub, and Adaptation Fund. Assess the resources and support required by CWRD and sector groups to operationalize this road map to accelerate mobilization of climate and green finance for adaptation, resilience, and low-carbon development.	• CWRD RMT • CCSD • Sector groups
F.	Increase investment in climate action derived from trust funds.	Trust funds contribute a small share of total operational finance in CWA. Despite prevailing high administrative hurdles and low funding amounts per investment, projects are encouraged to increase the share of trust fund finance, specifically for climate-related project preparation measures, analyses, and diagnostic studies.	• Sector groups • CWRD
G.	Raise more government contributions toward projects that address climate needs and priorities.	Develop innovative project financing solutions that combine concessional climate finance with leveraging DMCs' own sovereign contributions to finance projects: Azerbaijan, Georgia, Kazakhstan, Pakistan, Turkmenistan, and Uzbekistan.	• CWRD CMTs • Sector groups

ADB = Asian Development Bank, ADF = Asian Development Fund, CAREC = Central Asia Regional Economic Cooperation, CCSD = Climate Change and Sustainable Development Department, CMT = country management team, CWA = Central and West Asia, CWRD = Central and West Asia Department, DMC = development member country, GCF = Green Climate Fund, IF-CAP = Innovative Finance Facility for Climate in Asia and the Pacific, NDC = nationally determined contribution, OMDP = Office of Markets Development and Public-Private Partnership, OCR = ordinary capital resources, PBL = policy-based loan, PSOD = Private Sector Operations Department, RMT = regional management team.

[a] A PBL that aims at more than 80% of finance being attributed to ADB climate finance. Such PBLs can accommodate a diverse spectrum of climate action, not necessarily coherently aiming at adaptation or mitigation outcomes, but predominantly pursuing significant enhancement of a DMC's climate policy framework.

[b] The focus of thematic window (i) is on Type 2B climate change adaptation projects. These projects have adaptation as the primary objective and directly respond to the priorities identified in national climate change adaptation plans and disaster risk management plans in a programmatic manner to build resilience and minimize loss and damage (Type 2B climate change adaptation projects). The focus of thematic window (ii) is on topics such as regional health security, with attention to universal health coverage, food security, and environmental management such as ocean health and biodiversity and ecosystem services. Source: ADB. 2024. *Concessional Assistance Policy for the ADF-14 Period*.

Source: ADB.

78. Concessional financial resources need to be prioritized among CWA DMCs to ensure they deliver the most impact and catalyze other investments. To prioritize financial resources, a set of decision criteria are proposed: (i) consideration of the DMCs' climate mitigation or adaptation priorities based on need (see section Mobilizing Climate Finance), (ii) the readiness of private sector involvement to mobilize climate investment (see section Strategic Directions for Climate Action), and (iii) public debt levels eventually inhibiting sovereign borrowing for climate action. Table 21 illustrates the allocation of priorities among CWA DMCs. As Tajikistan and Pakistan stand out as priorities for adaptation financing through concessional finance and grants, and as Pakistan and Uzbekistan offer high potential that ADB concessional financing can mobilize private sector involvement, these three DMCs warrant particular focus. Pakistan and the Kyrgyz Republic both have high public debt and thus would be in even greater need of grant or concessional finance, provided that support does not reward high debt. The five other Group C DMCs would require concessional finance or grants from other financing facilities, such as from GCF. This is particularly important for action on adaptation, for which DMCs are more constrained to borrow.

79. ADB has a range of financing modalities suitable for financing specific needs to help DMCs meet their climate goals. Preparatory modalities can be deployed to properly prepare new projects and operations for climate objectives. Policy-based loans (PBLs) can be utilized to enhance climate policies at the national, regional, and sector level, and can support the preparation of regulatory frameworks to facilitate public and private climate finance. PBLs may be well suited for DMCs where there is readiness and appetite for enhanced climate policy, where climate change has not yet been fully integrated into sector and fiscal planning, and where policy reform could unlock potential for private sector participation in delivering investments needed for a resilient transition, such as in renewable energy, low-carbon transport, and sustainable urban development. Results-based loans can be used to tie loan disbursement to measurable climate results. For example, one of the targeted results could be a 20% reduction in economic losses due to climate hazards in the country by a specified date. Sector development programs can be used to couple policy reform with specific climate investments in a sector, for example, policy reform in the energy sector alongside investment to mobilize private sector involvement in delivering renewable energy generation. Appendix 4 provides a more detailed description of ADB financing modalities.

80. ADB will explore the potential to deliver to several DMCs climate PBLs specifically designed to formulate climate policies and achieve climate action. It may include financing to support the development of policy instruments and coordination mechanisms such as national climate change policies; the integration of climate goals into national development plans and annual budgets; financial systems and institutions for implementing, for example, a national sustainable finance strategy; pricing and subsidies, including the introduction of carbon taxes; and support for a just transition through, for example, vocational education and training. Climate PBLs can be applied to single or multisector programs, addressing underlying constraints to climate action and promoting greater investment for climate solutions, including through private sector investment and PPPs. ADB is currently preparing climate PBLs for the Kyrgyz Republic and Uzbekistan, and is exploring one for Kazakhstan.

Table 21: Prioritization of Climate Finance Resources to Deliver Climate Action

Categories	Afghanistan				Armenia				Azerbaijan				Georgia				Kazakhstan				Kyrgyz Republic				Pakistan				Tajikistan				Turkmenistan				Uzbekistan			
	A	M	PS	PD	A	M	PS	PD	A	M	PS	PD	A	M	PS	PD	A	M	PS	PD	A	M	PS	PD	A	M	PS	PD	A	M	PS	PD	A	M	PS	PD	A	M	PS	PD
Concessional assistance-only (ADF grants and concessional loans)																																								
OCR blend (concessional and market-based loans)																																								
OCR non-concessional (market based loans)																																								

Legend: Full color squares indicate high priority, lighter color moderate priority.

A = High social and human development vulnerability levels suggest adaptation investment should be prioritized to these countries.

M = The country was not on track as of 2022 to meet its nationally determined contribution target for 2030, suggesting investment in mitigation should be prioritized here.

PS = Successful mobilization of private climate finance in this country suggests investment could catalyze further impact.

PD = High public debt suggests concessional support would be highly needed in this country. Conditions need to be defined for this prioritization criterion to avoid rewarding high indebtedness. One option would be to invest with the condition that upstream policy support prepares bankable climate action plans, the use of capital and carbon markets leveraging additional financial resources, or the leveraging of additional climate cofinance.

ADF = Asian Development Fund, OCR = ordinary capital resources.

Source: Asian Development Bank.

4 IMPLEMENTATION, MONITORING, AND REPORTING

81. This section describes how *Climate Action Road Map for Central and West Asia, 2025–2030* shall be implemented, monitored, and reported within the institutional structures and strategic planning processes of ADB. It also describes how the road map shall be adjusted if necessary.

82. **Make climate finance a leadership priority within ADB.** Through its management structures at headquarters (director general and deputy director general) and its resident missions (country directors), ADB acknowledges its leadership responsibility to actively contribute to DMC climate policy dialogue, coordination structures, and institutional partnerships. ADB will identify and develop tailor-made solutions through the One ADB approach and collaborate with DMC government teams across ministries and state agencies at the national and subnational level. This will support ADB's acceptance as a partner and recognition as the climate bank for Asia and the Pacific by DMC governments and development partners.

83. **Integrate the road map into the country partnership strategy and programming framework of DMCs in CWA.** Country management teams shall require the alignment of the 5-year country partnership strategy (CPS) to the climate action road map for CWA in both processes—midterm review of the previous CPS and formulation of the new CPS. The road map will be accompanied by country-specific climate action plans to guide the annual country programming process. Country management teams will play a crucial role in monitoring progress in operationalizing the plan.

84. **Schedule a midterm review.** The climate action road map for CWA recommends specific targets for climate and adaptation finance volumes in the region. The objective of the review will be to adjust measures and actions if the achievement of targets appears to become unlikely. However, increased ambition should be considered and, if appropriate, implemented through raised targets.

85. The climate action road map for CWA is supported by both a capacity-building strategy and a communications strategy, each playing a crucial role in its implementation. The capacity-building strategy aims to systematically develop capacity for climate mitigation and adaptation investments in the DMCs of CWA. It provides a framework for long-term sector-specific, midterm institutional, and short-term individual capacity-building interventions at various levels: regional, subregional, and national. This approach emphasizes the importance of partnerships and coordinated efforts among development partners and national governments to ensure sustainable capacity development. Concurrently, the communications strategy serves as a guiding document for effectively leveraging communications to prepare stakeholders for the road map's full operationalization and capacity-building program. It focuses on guiding outreach and information dissemination among ADB, DMCs, development partners, and other stakeholders, such as the private sector and civil society organizations. Together, these strategies ensure a comprehensive approach to addressing climate change challenges in the region, fostering both capacity development and effective climate communication.

APPENDIX 1
CLIMATE IMPACTS ON SECTORS ACROSS CENTRAL AND WEST ASIA: KEY RISKS AND OPPORTUNITIES

A. Agriculture, Food, and Natural Resources

1. Agriculture and food security will be substantially affected by climate change, particularly in areas that depend on cereal production.[1] The main agricultural products across the 10 developing member countries (DMCs)[2] of the Asian Development Bank (ADB) are vegetables and fruits, while some countries grow significant amounts of cereal and fiber crops (Table A1.1). Livestock production is also significant, with approximately 600 million animals raised or slaughtered every year from 2000 to 2022. Crop and cattle production is highest in Pakistan, and crop yields are highest in Kazakhstan and Uzbekistan.

- **Physical risks.** Climate change will threaten food security and strain economic development. Heat increases crop and livestock water demand and growth stress, and affects agricultural workers' health and productivity.[3] Increases in crop water stress will lead to decreases in agricultural production and lower yields. More unpredictable and erratic water availability because of melting glaciers increases pressure on groundwater resources, which are already overexploited, particularly in Pakistan.[4] Alongside water stress, drought increases soil salinity, soil erosion, and desertification. At the other end of the spectrum, extreme precipitation and floods can cause waterlogging, harming crops and livestock and making agricultural land inaccessible.
- **Transition risks.** Greenhouse gas (GHG) emissions from agriculture are substantial, particularly in Pakistan. Emissions of nitrous oxide and methane continue to rise due to increased fertilizer application and increased production of livestock and crops.[5] The burning of crop residues causes severe air quality impacts alongside GHG emissions, particularly in Pakistan. Changes in agricultural policies to reduce GHG emissions pose potential risks to agricultural producers, as does consumer sentiment evolving toward greener diets and production methods. Countries with large portions of the population employed in agriculture are particularly vulnerable to shifts in labor market demand driven by climate change: notably, Armenia with 52% of the working population, Afghanistan with 45%, and Tajikistan with 44%.

---

1 R.Shaw et al. 2022. Chapter 10: Asia. In H.-O. Pörtner et al., eds. *Climate Change 2022: Impacts, Adaptation and Vulnerability.* Contribution of Working Group II to the Sixth Assessment Report of the Intergovernmental Panel on Climate Change. Cambridge University Press.

2 ADB placed its regular assistance to Afghanistan on hold effective 15 August 2021. This document covers Armenia, Azerbaijan, Georgia, Kazakhstan, the Kyrgyz Republic, Pakistan, Tajikistan, Turkmenistan, and Uzbekistan.

3 ADB. 2012. *Sector Briefing on Climate Change Impacts and Adaptation: Urban Development.*

4 T. Gleeson et al. 2012. Water Balance of Global Aquifers Revealed by Groundwater Footprint. *Nature.* 488. pp. 197–200.

5 J. P. Aryal et al. 2020. Climate Change Mitigation Options among Farmers in South Asia. *Environment, Development and Sustainability.* 22. pp. 3267–3289.

Table A1.1: Indicators of the Scale of Agricultural Production, Food Insecurity, Employment, and the Economic Value of Agriculture in Central and West Asia

Country	Gross Primary Crop Production in 2022 (million tons)	Total Meat Production in 2022 (million tons)[b]	Population Facing Food Insecurity (2020–2022) (%)[c]	Population Employed in Agriculture, 2022 (%)[d]	GDP from Agriculture in 2022 (%)[e]
Afghanistan	5.2	0.27	80	45	26
Armenia	3.9	0.10	<10	52	10
Azerbaijan	16.9	0.37	10–25	36	5
Georgia	2.8	0.07	25–40	40	6
Kazakhstan	76.0	1.24	<10	13	5
Kyrgyz Republic	10.9	0.26	<10	25	11
Pakistan	329.4	5.25	40–60	36	22
Tajikistan	13.0	0.38	No data	44	22
Turkmenistan	9.8	0.36	No data	23	12
Uzbekistan	64.0	1.35	25–40	26	24

Relative scale:

Very high High Medium Low

GDP = gross domestic product.

[a] Source: Food and Agricultural Organization of the United Nations (FAO). Crops and Livestock Products (accessed 8 August 2024).

[b] From cattle, chickens, goats, pigs, rabbits, and sheep. Source: FAO. 2024. Crops and Livestock Products (accessed 8 August 2024).

[c] Prevalence of moderate or severe food insecurity in the total population. Source: FAO. Prevalence of undernourishment (%) (3-year average) (accessed 8 August 2024).

[d] Source: International Labour Organization. ILOSTAT (accessed 8 August 2024).

[e] Source: World Bank national accounts data and Organisation for Economic Co-ordination and Development National Accounts data files.

- **Opportunities.** Major opportunities arise from regenerative agricultural practices, water-efficient technologies such as alternate wetting and drying for rice, sustainable land management that benefits biodiversity, enhancing carbon capture in soils and forests, and capturing methane and generating biogas from agricultural waste. Higher temperatures and increasing atmospheric carbon may increase agricultural productivity in some cases, shorten growing seasons, and widen the geographic suitability for agriculture by, for example, expanding it at higher elevations.[6] However, the long-term negative impacts from climate change generally outweigh the positives in most countries.

6 R. R. Wijngaard et al. 2018. Climate Change vs. Socio-Economic Development: Understanding the Future South Asian Water Gap. *Hydrology and Earth System Sciences*. 22. pp. 6297–6321.

B. Energy

2. Most of Central and West Asia relies heavily on fossil fuel energy generation (Table A1.2), and fossil fuel exports comprise a significant proportion of gross domestic product (GDP) in Azerbaijan, Kazakhstan, Turkmenistan, and Uzbekistan. Hydropower generation is significant in the Kyrgyz Republic, Pakistan, and Tajikistan. Renewable energy generation other than hydropower remains relatively small. Across all countries, transport remains a key sector using little energy from renewable sources.

Table A1.2: Energy Consumption by Source in Central and West Asia, 2022
('000 terajoules and % of primary energy)

Country	Oil	Coal	Gas	Hydro	Biofuels and waste	Nuclear	Wind & solar	Total
Afghanistan		159.6 81%		36.5 19%	No data	0	0.3 ~0%	196.4
Armenia	25 15.8%	0.5 0.3%	98 62.3%	8 5%	2 1.6%	22 13.9%	2 1.1%	157.2
Azerbaijan	227 32.5%	0	461 66.1%	5 0.7%	4 0.6%	0	1 0.1%	697.6
Georgia	58 26.7%	9 4.4%	101 46.8%	37 17%	10 4.6%	0	1 0.5%	215.6
Kazakhstan	717 24.9%	1,404 48.8%	710 24.7	33 1.2%	1 0.1%	0	12 0.4%	2,877.5
Kyrgyz Republic	17 36.9%	20 26.8%	24 8.4%	46 28%	<0.1%	0	0	107.3
Pakistan	1,014 19.9%	698 13.7%	1,481 29%	121 2.4%	1,610 31.5%	167 3.3%	15 0.3%	5,105.5
Tajikistan	46 28.4%	40 25%	8 5.2%	67 41.4%	0	0	<0.1%	162.2
Turkmenistan	253 24.8%	0	766 75.2%	<0.1%	<0.1%	0	0	1,018.7
Uzbekistan	161 7.9%	108 5.3%	1,756 85.9	18 0.9%	<0.1%	0	0	2,044.0

Relative scale:

Very high | High | Medium | Low

Source: International Energy Agency. 2024. *Country Energy Mix*; and LowCarbonPower. Electricity in Afghanistan in 2022.

- **Physical risks.** Heat reduces the efficiency of energy generation, transmission, and distribution and causes more frequent failure of systems and components.[7] In addition, increased water temperature hampers cooling in thermal and nuclear power stations. More erratic flows may render cooling water unavailable. Efficiency and generation in hydropower plants will be affected as higher glacier melt increases water availability in the near term, but reduces

[7] For a country with 8% line losses, a 3°C temperature increase will cause an increased need for generation of about 1%. Source: J. F. Feenstra et al. 1998. *Handbook on Methods for Climate Change Impact Assessment and Adaptation Strategies*. United Nations Environment Programme.

in the second half of the century.[8] More extreme flows will increase sedimentation in rivers, shortening the lifetime of hydropower infrastructure. Energy infrastructure and transmission face risks from climate hazards, including floods and landslides. More frequent heat will increase energy demand for cooling purposes. A 1°C degree increase in ambient temperature can raise demand for electricity by 0.5%–8.5%.[9] A strain on energy networks was evident in the 2018 heat wave affecting Armenia and Azerbaijan.[10]

- **Transition risks.** GHG emissions from the energy sector are particularly high in Kazakhstan and Pakistan, and all DMCs rely on fossil fuels to some degree—more so for heating and transport than for electricity in some DMCs. Changes in energy policy and technologies pose risks to existing energy generators and could increase energy prices in the short to medium term. The huge increase in demand for electricity brings challenges to balancing supply and demand, particularly with more decentralized renewable generation. Several DMCs rely on imported energy, including electricity, notably Afghanistan, which imports over 80% of its electricity. This creates risks relating to energy security and affordability with the transition to low-carbon energy.
- **Opportunities.** Considerable opportunities exist for increased energy efficiency, demand reduction, and renewable energy generation—in particular hydropower, solar, wind, geothermal, and bioenergy. However, sustainable and resilient land and water use will be critical to unlock this potential. In addition, there are opportunities in hydrogen, carbon capture and storage, and decarbonizing heat, as through electric heat pumps. Climate warming may reduce energy requirements for heating purposes.

C. Transport

3. Most transport is by road in Central and West Asia, as few countries have significant or countrywide rail networks (Table A1.3). As most of the countries are landlocked, resilient regional road connections are of major importance for the trade and supply of essential goods. Increased urbanization rates will require improvements to the efficiency and sustainability of urban transport systems to meet the increased mobility demands and support urban economic activities.

8 In the Caucasian region, this tipping point is likely already reached or will be reached within the coming decades. In the Central Asian region and the mountains of Pakistan, this point will be reached toward the end of the century. Source: M. Huss and R. Hock. 2018. Global-Scale Hydrological Response to Future Glacier Mass Loss. *Nature Climate Change*. 8. pp. 135–140.

9 M. Santamouris et al. 2015. On the Impact of Urban Heat Island and Global Warming on the Power Demand and Electricity Consumption of Buildings—A Review. *Energy and Buildings*. 98. pp. 119–124.

10 World Bank Group and ADB. 2021. *Climate Risk Country Profile: Armenia*.

Table A1.3: Total Road Network Length and Average Passenger Kilometers per Year by Rail and Other Means in Central and West Asia, 2018–2021

Country	Length of Road Network (kilometers)	Average Travel by Rail (kilometers per person per year)	Average Travel by Road[a] (million kilometers per person per year)
Afghanistan	25,907	No data	No data
Armenia	30,262	195.3	1,180 (2023)
Azerbaijan	76,808	1,286.0	21,119 (2023)
Georgia	102,855	1,830.9	7,551 (2022)
Kazakhstan	383,086	58,196.5	91,298 (2020)
Kyrgyz Republic	34,000	99.9	5,691 (2017)
Pakistan	1,041,331	74,984.0	325,000 (2010)
Tajikistan	14,140	0.5	10,600 (2015)
Turkmenistan	13,773	1.1	1,065 (n.a.)
Uzbekistan	178,390	13,639.3	144,700 (2022)

Relative scale:

Very high High Medium Low

n.a. = not available.

Sources: First two columns, CEIC Data; and third column, International Road Federation. 2024. *Road Network.*

- **Physical risks.** Heat stress damages roads, pavements, rail lines, and airport runways (footnote 3). People using public and private transport during heat waves can be more exposed to heat risks from, for example, waiting for the bus without shade or crowding on hot trains. In the region, several important roads lead through mountain ranges as economic corridors. This infrastructure is at risk from more floods, landslides, rockfalls, avalanches, mudflows, and glacial lake outbursts. Additionally, increased heavy rainfall can flood roads, particularly in urban areas. Air quality impacts are a major concern from transport, particularly in cities, and this can be exacerbated during heat waves, when ozone generation can increase, and during days with low wind. Supply and distribution chains are likely to become more vulnerable to disruption within and outside the region. Maritime shipping, which accounts for around 80% of global trade by volume, could experience more frequent disruptions.[11] This could affect ports at Baku, Azerbaijan; Batumi, Georgia; Karachi, Pakistan; and Turkmenbashi, Turkmenistan.
- **Transition risks.** Fuel combustion in transport is one of the largest sources of GHG emissions, and the sector faces increasing exposure to carbon pricing initiatives globally.[12] Changes in policies affecting internal combustion vehicles, including policies to improve air quality, may increase the price of transport services and vehicles. The increase in electric vehicles will place heavy demand on electricity systems, and this needs to be carefully managed to balance energy security and consumer confidence in e-mobility. The supply of batteries for electric vehicles is at risk from, for example, reliance on rare earth metals from fragile countries, as is the recycling and disposal of batteries.

[11] R. Dellink et al. 2017. International Trade Consequences of Climate Change. *OECD Trade and Environment Working Papers.* No. 2017/01. OECD.

[12] United Nations Environment Programme. 2024. *Climate Risks in the Transportation Sector.*

- **Opportunities.** Improvements to transport efficiency and the transition to electric- and hydrogen-powered vehicles present opportunities, alongside an increase in mode shift to public transport, cycling, and walking. Uzbekistan is notable, among others, for its potential to build up its own manufacturing industry for electric vehicles, servicing the region's demand at reasonable transaction costs.

D. Water and Sanitation

4. Water security is a major priority across the region, with Pakistan, Turkmenistan, and Uzbekistan facing the greatest risks as high water demand exceeds water availability under climate change (Table A1.4). In some regions, overexploitation of surface and groundwater is deteriorating water quality, depleting aquifers, and causing saline intrusion. Sustainable management of major river basins, such as the Indus Basin in Pakistan, plays a critical role in social and economic development, and presents possible areas for tension over water access. Safe sanitation and wastewater treatment varies across DMCs and is an area of major need to address health and environmental sustainability goals.

Table A1.4: Projected Water Stress Index in ADB Developing Member Countries in Central and West Asia

Country	Agriculture	Domestic	Industry	Average across the Three Sectors
Afghanistan	3.4	2.5	2.3	3.4
Armenia	3.0	3.2	3.2	3.1
Azerbaijan	2.4	3.0	2.9	2.7
Georgia	1.5	1.1	1.0	1.2
Kazakhstan	3.0	2.4	2.6	2.8
Kyrgyz Republic	3.5	3.3	3.4	3.5
Pakistan	3.9	3.7	3.7	3.9
Tajikistan	2.5	2.5	2.7	2.5
Turkmenistan	3.7	3.5	3.8	3.8
Uzbekistan	3.8	3.6	3.8	3.8

Relative scale:

Very high High Medium Low

Note: Projected Water Stress Index is determined assuming a high climate change scenario (RCP8.5) for the agriculture, domestic, and industry sectors in the 2050s. Water stress is indicated on a scale from 0 to 5, assessing competition for water resources, and is defined as the ratio of demand for water by human society divided by available water. The lower the ranking number, the higher the aggregated water stress risk.

Source: Aquedact. Country Rankings (accessed 8 August 2024).

- **Physical risks.** More frequent and longer periods of heat accelerate evaporation rates and increase water demand for agriculture and domestic supply. In Central Asia, a 30% decrease in glacier areas over the last 50 years has reduced river flows even as water demand continues to increase.[13] Heat can reduce the performance of wastewater treatment facilities. Increasing hydrological variability and periods of reduced rainfall will significantly worsen water scarcity. A study in Central Asia suggests that, under a business-as-usual scenario, a combination of increasing demand and decreasing supply would widen the water supply–demand gap from a current 8% to 37%.[14] More frequent heavy precipitation, floods, and debris flows affect water resource management infrastructure such as reservoirs and diversion works. In addition, increasing sedimentation rates infill reservoirs more rapidly, reducing their storage capacity. Increased rainfall can worsen runoff from agriculture and pollute sources, affecting water quality and drinking water safety, partly from unsafe sanitation and overflowing sewers.
- **Transition risks.** Water resource management can be energy intensive, requiring, for example, pumping to convey water. However, the greatest impact on GHG emissions in this sector comes from wastewater treatment or the lack of it. Policies on sustainable waste management and pollution reduction pose some risks to infrastructure owners but bring many benefits for environmental protection.
- **Opportunities.** Enhancing water efficiency, rainwater reuse schemes, and in some cases low-energy desalination are key priorities. Nature-based solutions to reduce water pollution and sustainable wastewater treatment are also required.

E. Urban Development

5. Urbanization is intensifying in most countries, generally increasing demand for energy and water, as well as the number of people exposed to physical climate risks in urban settlements. For example, the urban population in Pakistan is growing by approximately 3% every year, and by 2030 it is expected that more than half of Pakistan's population will live in cities.[15] One in eight urban dwellers lives below the poverty line for lack of public services and productive jobs. Table A1.5 highlights the climate risk scores of the largest city in each of the DMCs, indicating major risks for Karachi, Pakistan and Dushanbe, Tajikistan.

- **Physical risks.** Climate change significantly complicates urban planning by intensifying issues such as flooding, partly from extensive paved surfaces, urban heat, and the effects of extreme weather on essential infrastructure.[16] For example, Tashkent, Uzbekistan experiences a significant urban heat island affecting nearly 25% of its area. In some cities, mitigation of this problem with urban greening solutions is difficult to realize. Ashgabat, Turkmenistan, for example, faces extreme heat but also water stress.[17] Cities also face risks from water shortages and poor air quality exacerbated by climate change. Cities on the coast of Pakistan face risks from sea level rise, cyclones, and storm surges.

---

13 A. Haidar. 2023. EDB Report Warns of Looming Water Crisis in Central Asia by 2028. *Astana Times*. 22 December.

14 J. Hunink and J. Bird. 2021. *Scoping Study for the CAREC Water Pillar: Climate Resilience Through Regional Cooperation*. ADB.

15 Consortium for Development Policy Research. The Six Biggest Challenges Facing Pakistan's Urban Future.

16 ADB. 2012. *Guidelines for Climate Proofing Investment in Agriculture, Rural Development, and Food Security*.

17 N. Koch. 2015. The Violence of Spectacle: Statist Schemes to Green the Desert and Constructing Astana and Ashgabat as Urban Oases. *Social & Cultural Geography*. 16 (6). pp. 675–697.

Table A1.5: Climate Risk Ratings of the Largest Cities in Central and West Asia

Country	Largest City	Population (2024 estimated)[a]	Relative Climate Risk Rating
Afghanistan	Kabul	4,434,550	High
Armenia	Yerevan	1,097,542	Medium
Azerbaijan	Baku	2,464,162	Low
Georgia	Tbilisi	1,084,471	Medium
Kazakhstan	Almaty	2,015,209	Low
Kyrgyz Republic	Bishkek	1,127,721	Medium
Pakistan	Karachi	17,648,555	Very high[a]
Tajikistan	Dushanbe	1,012,794	High
Turkmenistan	Ashgabat	921,601	Low
Uzbekistan	Tashkent	2,633,661	Low

Relative scale:

Very high High Medium Low

[a] Karachi scored highest among the largest cities in the region concerning the climate risk and hazard assessment. See ADB. 2017. *Climate Change Profile of Pakistan*. Appendix 6.

Sources: World Population Review. World Population by Country 2024 and Karachi Population 2024 (for population) (accessed 1 July 2024); and Think Hazard! (accessed 1 July 2024) and EU Inform Climate Risk Assessment (for risk rating).

- **Transition risks.** Urban settlements are major GHG emitters due to consumption of energy, materials and goods, and, in particular, buildings and transport. The increase in air-conditioning of buildings poses large risks to GHG emissions and electricity load balancing, and in some cases threatens energy security alongside a surge in electricity demand required to power electric vehicles. In more affluent countries, urban lifestyles are linked to high-profit sectors which can also be high GHG emitters: fossil fuels, industry, and mining. Therefore, as the economy in some countries decarbonizes, there are various risks to urban sectors that depend on wealth stemming from high-emitting sectors, notably retail and hospitality, which appeal to wealthier urban populations.
- **Opportunities.** While climate change may offer opportunities for urban expansion to higher altitudes, this is site-specific and comes with increased mountain hazards. Adapting urban planning to enhance resilience could promote employment in climate action. Cities have significant opportunities with urban greening, public transport and active travel, decentralized renewables such as rooftop solar, decarbonizing buildings and housing, and local actions to reduce hotspots for GHG emissions and climate risks, such as enhancing sustainable solid waste management and hazard warning systems.

F. Health

6. Across the region, countries are facing health issues arising from malnutrition, air pollution, water pollution, poor hygiene, infectious diseases, and chronic illnesses. Climate change exacerbates existing physical and mental health conditions, particularly affecting older people, the unwell, people with disabilities, pregnant women, and infants (Table A1.6). Health indicators among the 10 DMCs underline that Afghanistan and Pakistan in particular have severe needs.

Table A1.6: Selected Health Indicators in Central and West Asia

Country	Life Expectancy (years)	Prevalence of Stunting in Children under 5 Years Old (%)	Air Pollution Mortality Rate (per 100,000)	Exposure to Unsafe WASH Mortality Rate (per 100,000)	Change in Heat-Related Mortality Under 2°C Warming[a]
Afghanistan	62.9	44.7	177.2	16.6	5,000–10,000
Armenia	73.0	7.2	74.5	5.8	0–5,000
Azerbaijan	72.9	13.3	125.2	3.6	5,000–10,000
Georgia	71.2	4.8	92.8	3.3	0–5,000
Kazakhstan	70.3	4.9	83.4	3.2	5,000–10,000
Kyrgyz Republic	72.2	10.3	124.9	2.3	0–5,000
Pakistan	66.0	34.0	192.1	38.8	35,000–50,000
Tajikistan	71.8	13.1	203.8	9.0	0–5,000
Turkmenistan	69.1	6.7	87.9	5.7	5,000–10,000
Uzbekistan	72.2	6.9	151.7	2.9	5,000–10,000

Relative scale:

Very high High Medium Low

WASH = water, sanitation, and hygiene.

[a] Estimated absolute change in heat health-related deaths in adults over age 65 at mid-century (2041–2060) with respect to a 1995–2004 baseline, under 2°C warming.

Source: World Health Organization. World Health Statistics 2024.

- **Physical risks.** Heat stress is a major health concern from climate change, as it can cause heat stroke and respiratory and cardiovascular disease, as well as significantly distress people with mental health conditions. Higher temperatures can worsen drinking water quality, increase the spread of infectious diseases, affect food production and storage and so malnutrition, and exacerbate air quality issues. Heat stress also has major impacts on labor force productivity. Research suggests that global labor productivity during the hottest months has already dropped by 10% because of warming, and that a decline of up to 20% may be expected by 2050 under high emissions. More frequent drought may hamper water availability for drinking and affect water quality, hygiene, and sanitation. For example, poor quality drinking water in the lower parts of the Amu Darya has been linked to increases in kidney, thyroid, and liver diseases.[18] Vulnerable people who depend on surface water for drinking are affected most.

[18] GRID-Arendal. 2011. *Environment and Security in the Amu Darya Basin.*

Increased flooding and overflowing of sewers can spread infectious disease. In addition, trauma and injury resulting from floods, landslides, and wildfires can significantly undermine health and well-being.

- **Transition risks.** The health sector is not highly exposed to transition risks, as GHG emissions from hospitals and care homes are relatively small. However, supply chains for medicines and other medical supplies may be impaired by changes to industrial and transport policies.
- **Opportunities.** Major opportunities arise from enhancements to health systems, investment in early warning systems and preventative measures, and nature-based solutions to manage heat and support positive physical and mental health.

G. Private Sector and Finance

7. The private sector is very diverse across the region. It comprises companies focused on mining, heavy and light industries such as chemicals and manufacturing, and construction, all of which can have significant GHG emissions. Fuel, electricity, and materials used on-site and in manufacturing processes are major contributors to GHG emissions from the private sector. The finance sector plays a key role in enabling investment in the private sector to facilitate climate mitigation and adaptation by, for example, lending to smaller enterprises and smallholder farms for investment to enhance climate resilience. Table A1.7 presents a snapshot of the private sector landscape across the 10 countries. This is accompanied with scores from the World Economic Forum's Global Competitiveness Index, providing an indication of economic competitiveness and productivity, and scores from the European Bank for Reconstruction and Development Financial Market Development Index, which indicates the extent of favorable conditions for financial markets and the depth and maturity of financial markets. The indicators show Armenia, Azerbaijan, Georgia, and Kazakhstan as the most mature in terms of financial markets and private sector competitiveness.

Table A1.7: Private Sector Landscape, Economic Competitiveness, and Financial Market Development

Country	Private Sector Landscape	Global Competitiveness Index 2019[a]	Financial Market Development Index 2021[b]
Afghanistan	Domestic credit to the private sector was 3.07% of GDP in 2020.[c] The banking system remains severely constrained, limiting the availability of credit for private businesses. Many firms rely on the informal *hawala* system for payments.	Not scored	Not scored
Armenia	Domestic credit to the private sector equaled 55.4% of GDP in 2023.[d] Access to finance remains a key constraint to private businesses, and high interest rates remain a deterrent to investment.[e]	61.3 (Rank 69)	44.1
Azerbaijan	Domestic credit to the private sector was 23.2% of GDP in 2023. The private sector share of GDP was more than 80% in 2018.[f] Private sector activity tends to be concentrated in Baku and the Abseron Peninsula.	62.7 (Rank 58)	30.1
Georgia	The private sector contributed 86% of GDP and 81% of formal employment in 2022.[g] Shortages of key skills and poor learning outcomes are key barriers to private sector growth.	60.6 (Rank 74)	38.4

Country	Private Sector Landscape	Global Competitiveness Index 2019[a]	Financial Market Development Index 2021[b]
Kazakhstan	Domestic credit to the private sector was 26.1% of GDP in 2023. The electricity sector aims to stimulate investment in aging heat and power infrastructure. Extractive-fueled growth has failed to drive private sector growth, showing a need to diversify the private sector.[h]	62.9 (Rank 55)	47.3
Kyrgyz Republic	The private sector generates 70% of GDP, but low competitiveness and stagnant productivity have kept this unchanged since the 2000s.[i] Lack of access to finance, skilled labor, and regional trade links are key barriers.	54.0 (Rank 96)	34.2
Pakistan	Pakistan has relatively low private investment, which declined from the equivalent of 13.7% of GDP in 2000 to 10.0% in 2020.[j] Women tend to be underrepresented in the formal private sector.	51.4 (Rank 110)	Not scored
Tajikistan	The private sector generated 62% of GDP in 2017, concentrated in the production of goods, agriculture, trade, hospitality, and manufacturing.[k] Key barriers include low business entry rates as firms tend to invest little.	52.4 (Rank 104)	26
Turkmenistan	Although 60% of GDP was reported as private sector revenue, the true share is believed to be lower.[l] Strong state control over the private sector has slowed growth.	Not scored	0.9
Uzbekistan	Small and medium-sized enterprises account for >50% of GDP. Large state-owned enterprises dominate strategic sectors, limiting private sector growth.[m]	Not scored	27.7

Relative scale (inverse):

Very high High Medium Low

GDP = gross domestic product.

[a] Score out of 100. World Bank Group. WEF Global Competitiveness Index.

[b] Score out of 100. 2021. European Bank for Reconstruction and Development (EBRD). 2021. Chapter 5: The Financial Market Development Index. In *Transition Report 2021–22. System Upgrade: Delivering the Digital Dividend.* pp. 110–121.

[c] Trading Economics. Afghanistan—Domestic Credit To Private Sector (% of GDP) (accessed 1 August 2024).

[d] EBRD. 2024. *The EBRD in Armenia. Results Snapshot.*

[e] ADB. 2014. *Country Partnership Strategy: Azerbaijan, 2014–2018*.

[f] World Bank Group. Data. Domestic Credit to Private Sector (% of GDP) – Azerbaijan (accessed 1 August 2024).

[g] EBRD. 2023. *Kazakhstan Transition Report 2023–24*.

[h] Organisation for Economic Co-operation and Development. 2023. *Insights on the Business Climate in Kazakhstan*. OECD Publishing.

[i] EBRD. 2023. *Kygrz Republic Transition Report 2023–24*.

[j] World Bank Group. 2023. *ransforming Pakistan's Private Sector. Facilitating a Dynamic, Productive, and Environmentally Sustainable Economy. Policy Note 3*.

[k] D. Skakova and E. Livny. 2020. *Tajikistan Diagnostic*. EBRD.

[l] EBRD. Turkmenistan: Overview.

[m] EBRD. 2005. *Private Sector Assessment for Uzbekistan*.

Source: Compiled by the Asian Development Bank.

- **Physical risks.** Private sector operations face a range of risks from climate change, including damage to assets from extreme weather, insufficient water availability for operations, and stranded assets as locations become unfeasible for operations. The finance sector may be exposed to risks relating to investments that are not resilient or fail from flood damage, in particular investment in housing. These risks are rarely monitored or modeled for the future.
- **Transition risks.** The private and finance sectors can be at risk from abrupt changes to climate policies, technology change, and changes in market sentiment and preferences. High-emitting firms are particularly vulnerable to transition risks and are likely to face higher borrowing and capital costs in the medium to long term. Financial institutions face increasing pressure to support decarbonization carried out by their clients and to align their investment portfolios with climate action.
- **Opportunities:** The private sector has opportunities to enhance the resilience of operations and supply chains, enhance energy and water efficiency, implement low-carbon technologies, and reduce waste by, for example, generating biogas from industrial waste. Domestic financial institutions have major opportunities to finance climate action by, for example, providing green mortgages and sustainability-linked loans, but will require capacity building and supportive policies for this potential to be realized.

APPENDIX 2
ADB CLIMATE PORTFOLIO PER COUNTRY IN CENTRAL AND WEST ASIA, 2019–2023

8. Figures A2.1–A2.18 present Asian Development Bank (ADB) climate portfolio data based on project commitments for the 5-year period 2019–2023 for nine developing member countries in Central and West Asia.[19] ADB had no operations in Turkmenistan during these 5 years and Azerbaijan and Kazakhstan had mostly private sector operations. Total climate finance for each country is shown alongside its climate finance by sector.

Figure A2.1: ADB Climate Finance in Afghanistan, 2019–2023
($ million)

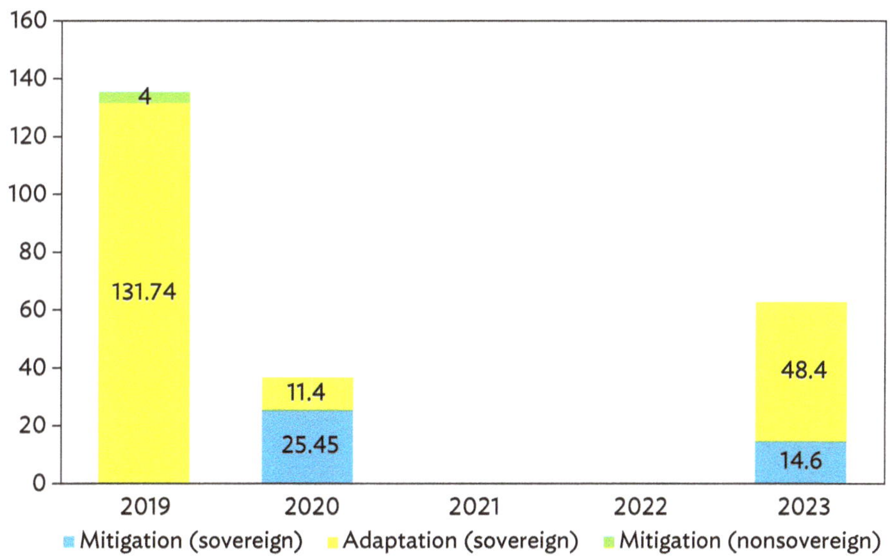

ADB = Asian Development Bank.
Source: ADB. Climate Finance Database (internal) (accessed 26 February 2024).

[19] ADB placed its regular assistance to Afghanistan on hold effective 15 August 2021. This document covers Armenia, Azerbaijan, Georgia, Kazakhstan, the Kyrgyz Republic, Pakistan, Tajikistan, Turkmenistan, and Uzbekistan.

Figure A2.2: ADB Climate Finance in Afghanistan by Sector, 2019–2023

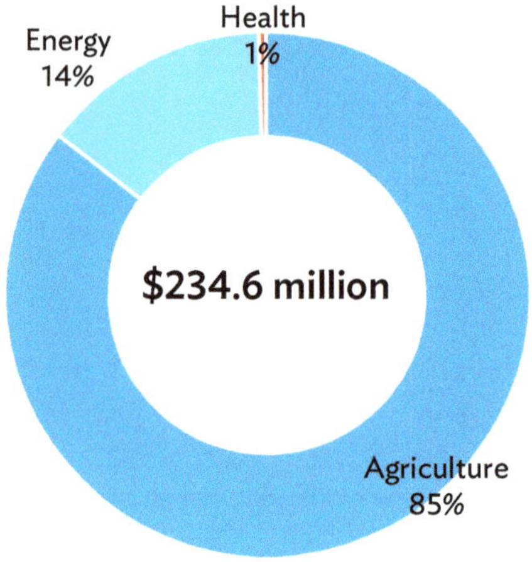

ADB = Asian Development Bank.
Source: ADB. Climate Finance Database (internal) (accessed 26 February 2024).

Figure A2.3: ADB Climate Finance in Armenia, 2019–2023
($ million)

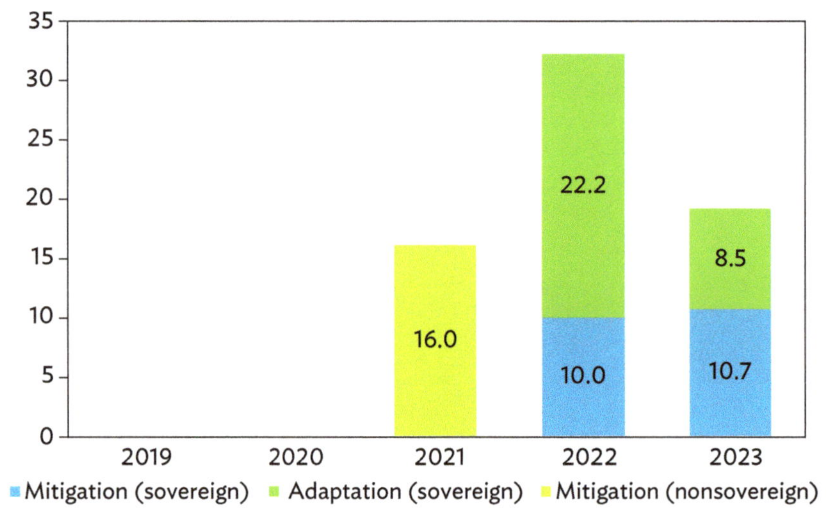

ADB = Asian Development Bank.
Source: ADB. Climate Finance Database (internal) (accessed 26 February 2024).

Figure A2.4: ADB Climate Finance in Armenia by Sector, 2019–2023

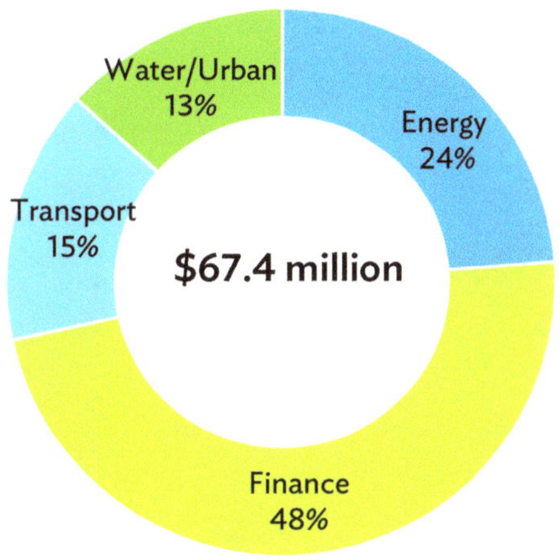

ADB = Asian Development Bank.

Source: ADB. Climate Finance Database (internal) (accessed 26 February 2024).

Figure A2.5: ADB Climate Finance in Azerbaijan, 2019–2023

($ million)

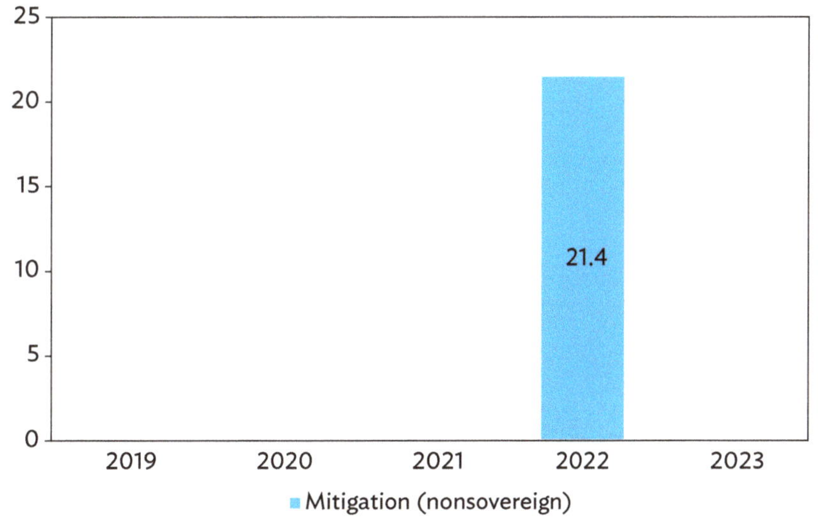

ADB = Asian Development Bank.

Source: ADB. Climate Finance Database (internal) (accessed 26 February 2024).

Figure A2.6: ADB Climate Finance in Azerbaijan by Sector, 2019–2023

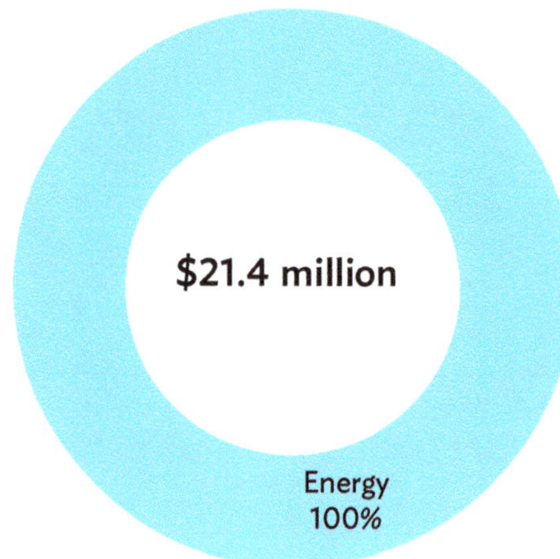

ADB = Asian Development Bank.
Source: ADB. Climate Finance Database (internal) (accessed 26 February 2024).

Figure A2.7: ADB Climate Finance in Georgia, 2019–2023
($ million)

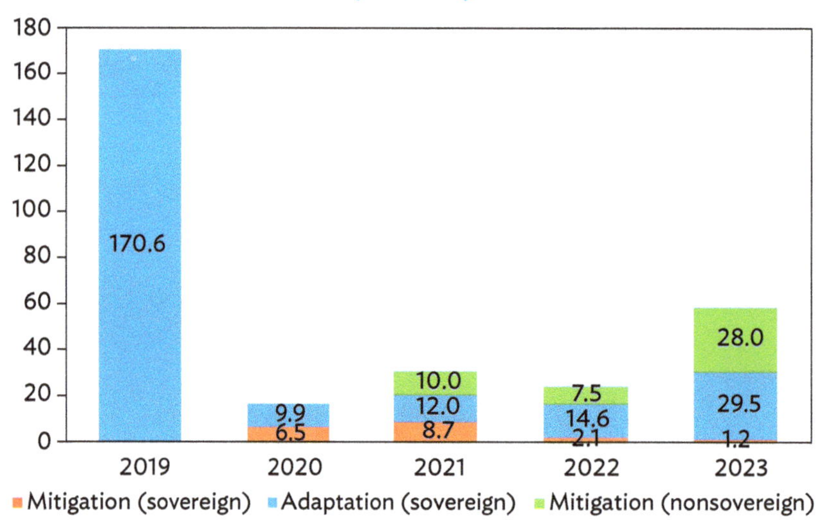

ADB = Asian Development Bank.
Source: ADB. Climate Finance Database (internal) (accessed 26 February 2024).

Figure A2.8: ADB Climate Finance in Georgia by Sector, 2019–2023

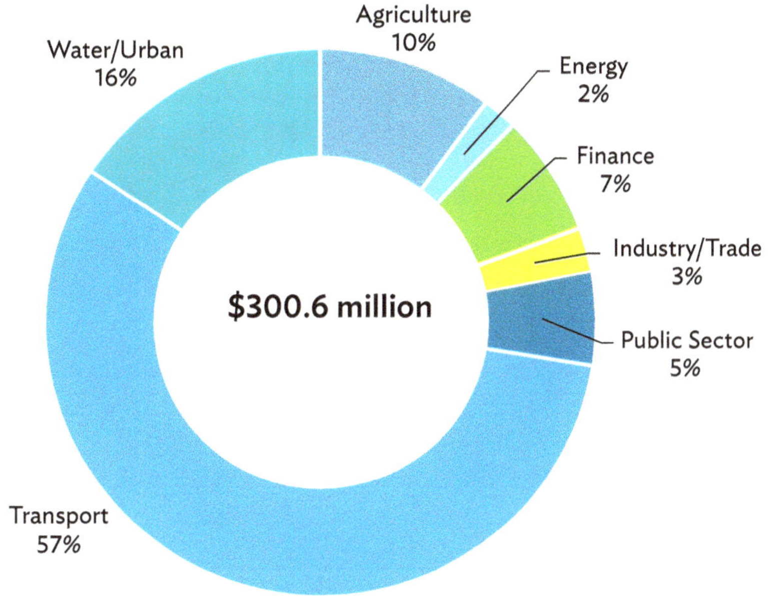

ADB = Asian Development Bank.
Source: ADB. Climate Finance Database (internal) (accessed 26 February 2024).

Figure A2.9: ADB Climate Finance in Kazakhstan, 2019–2023
($ million)

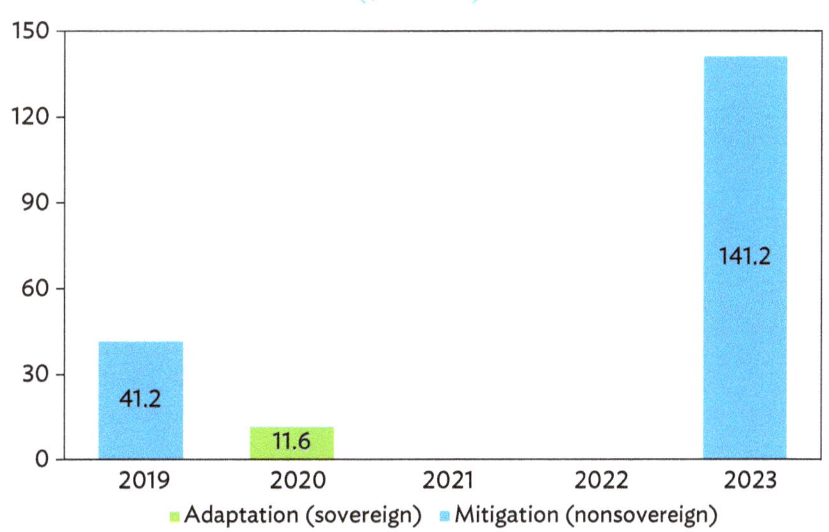

ADB = Asian Development Bank.
Source: ADB. Climate Finance Database (internal) (accessed 26 February 2024).

Figure A2.10: ADB Climate Finance in Kazakhstan by Sector, 2019–2023

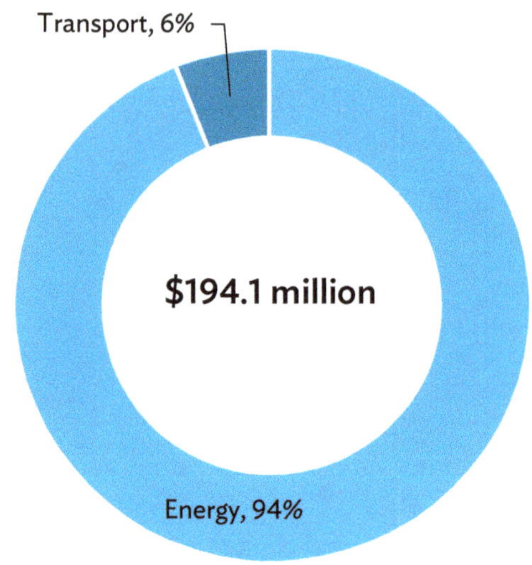

ADB = Asian Development Bank.

Source: ADB. Climate Finance Database (internal) (accessed 26 February 2024).

Figure A2.11: ADB Climate Finance in the Kyrgyz Republic, 2019–2023
($ million)

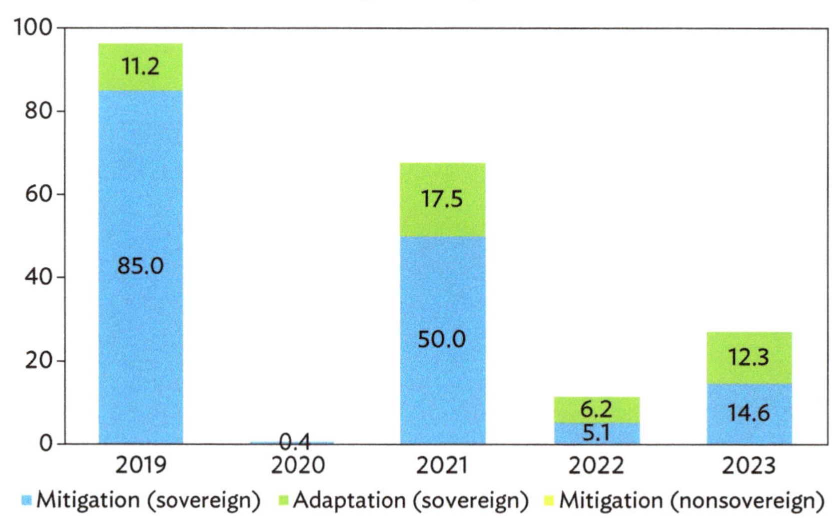

ADB = Asian Development Bank.

Source: ADB. Climate Finance Database (internal) (accessed 26 February 2024).

Figure A2.12: ADB Climate Finance in the Kyrgyz Republic by Sector, 2019–2023

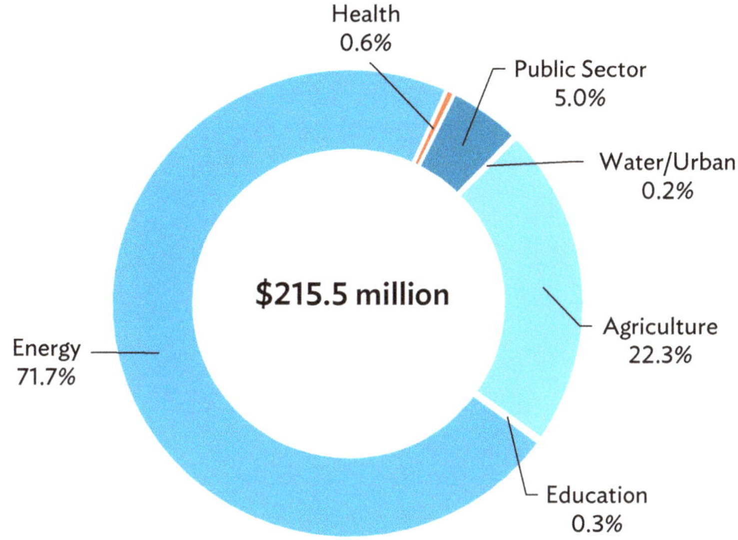

ADB = Asian Development Bank.
Source: ADB. Climate Finance Database (internal) (accessed 26 February 2024).

Figure A2.13: ADB Climate Finance in Pakistan, 2019–2023
($ million)

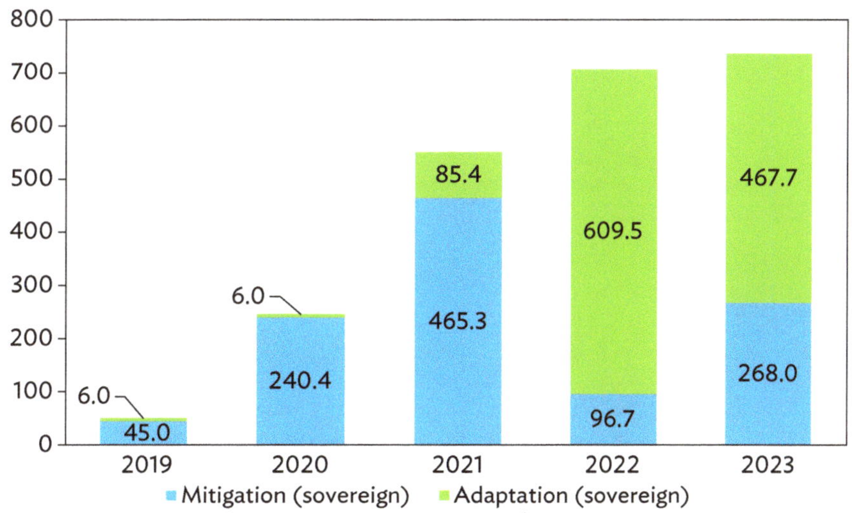

ADB = Asian Development Bank.
Source: ADB. Climate Finance Database (internal) (accessed 26 February 2024).

Figure A2.14: ADB Climate Finance in Pakistan by Sector, 2019–2023

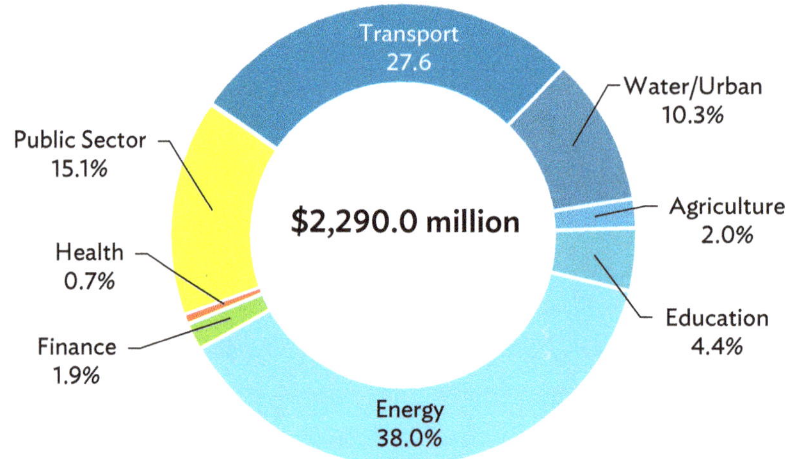

ADB = Asian Development Bank.

Source: ADB. Climate Finance Database (internal) (accessed 26 February 2024).

Figure A2.15: ADB Climate Finance in Tajikistan, 2019–2023
($ million)

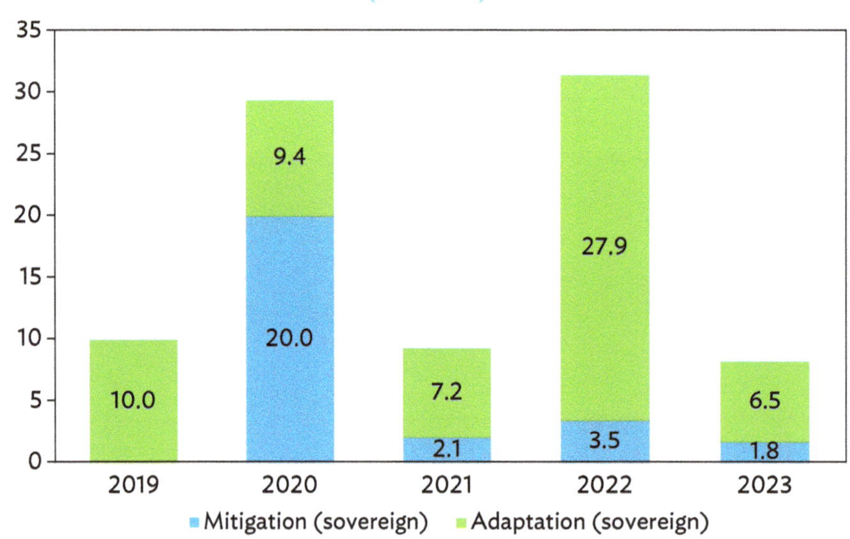

ADB = Asian Development Bank.

Source: ADB. Climate Finance Database (internal) (accessed 26 February 2024).

Figure A2.16: ADB Climate Finance in Tajikistan by Sector, 2019–2023

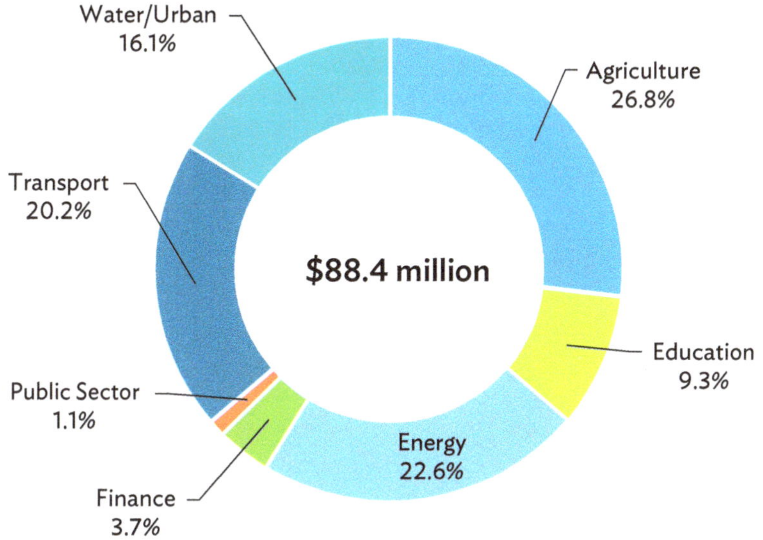

ADB = Asian Development Bank.
Source: ADB. Climate Finance Database (internal) (accessed 26 February 2024).

Figure A2.17: ADB Climate Finance in Uzbekistan, 2019–2023
($ million)

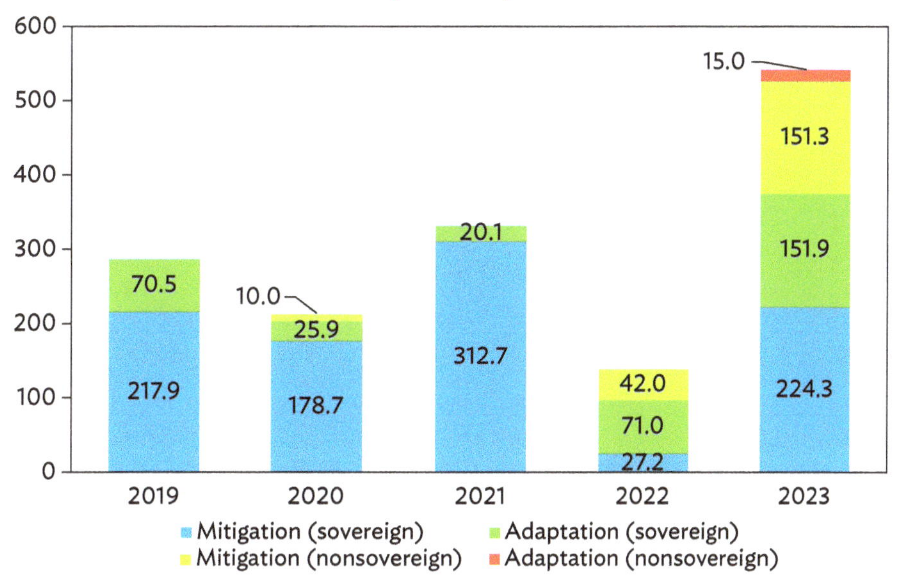

ADB = Asian Development Bank
Source: ADB. Climate Finance Database (internal) (accessed 26 February 2024).

Figure A2.18: ADB Climate Finance in Uzbekistan by Sector, 2019–2023

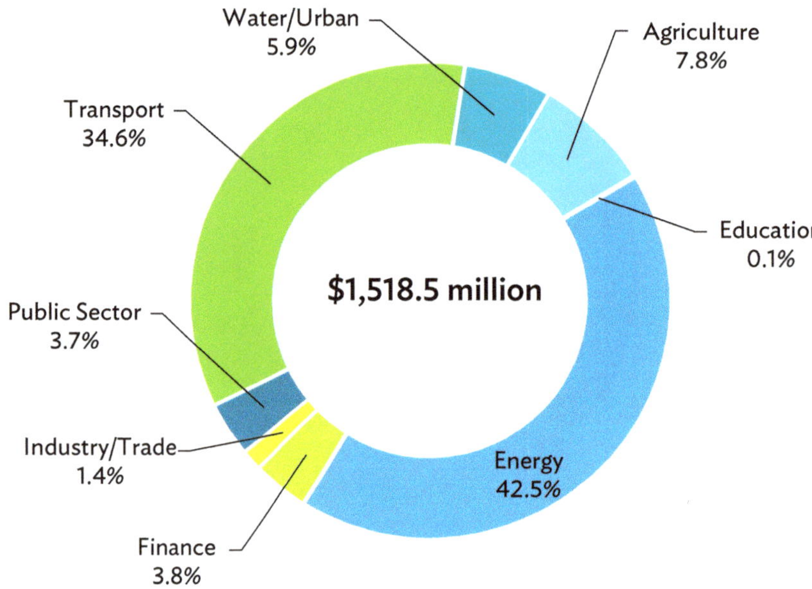

ADB = Asian Development Bank.

Source: ADB. Climate Finance Database (internal) (accessed 26 February 2024).

APPENDIX 3
EXAMPLES OF INTERCONNECTEDNESS OF ADB UPSTREAM, MIDSTREAM, AND DOWNSTREAM INTERVENTIONS

Objective	Upstream	Midstream	Downstream
Promoting green and sustainable finance development	Policy-based investment to promote green and sustainable finance frameworks (taxonomy and corporate and finance institution ESG guidelines)	TA with capacity building for SOEs, corporations, smaller enterprises, and finance institutions on ESG finance and ESG reporting and disclosure requirements through suitable anchor agencies	Commercial loans for concessional nonsovereign lending to sustainable and green finance intermediaries
Crowding in private sector investment for public services and utilities	Policy-based investment to improve the regulatory framework for power purchase agreements, feed-in tariffs, and off-taker arrangements; de-risking mechanisms, such as sovereign first-loss guarantees, for renewable energy investors	Preparation of investment projects and tendering processes for renewable energy infrastructure construction and operation concessions; digitization of public services and utilities management	Preparing private renewable energy investors for participation in bidding processes and financing of renewable energy investments; providing sovereign guarantees for investors
Transitioning to a low-carbon economy	Policy-based investment to support the development and implementation of green transition policies and long-term low-emission development strategies	TA for the development of decarbonization action and investment plans for high-emitting SOEs	Commercial or concessional loans for infrastructure investment by high-emitting SOEs to transition business processes to low-carbon technologies for cement, aluminum, steel, and fertilizers production
Developing climate and disaster insurance solutions	TA on innovative risk-transfer and finance solutions, such as regional insurance pools and sovereign catastrophe bonds	Policy-based investment for the enhancement of the regulatory framework of the insurance sector and targeted provision of sovereign co-funding for insurance premiums	Commercial or concessional loans, guarantees to underwriters, and loss-adjusters for scaling up successfully piloted products

ADB = Asian Development Bank; ESG = environmental, social, and governance; SOE = state-owned enterprise; TA = technical assistance.
Source: ADB.

APPENDIX 4
OVERVIEW OF ADB SOVEREIGN AND NONSOVEREIGN FINANCING MODALITIES

Investment Lending Modalities	
Project loan and grant	ADB's most deployed modality. It typically supports investments with a clear scope, tangible outputs, and estimated costs for goods, works, and services needed to complete the project. ADB is highly involved and uses its own procedures to prepare and administer project loans. This is most suitable for projects that need capacity building and activities that may pose adverse environmental and/or social impacts.
Sector loan	Finances numerous, smaller subprojects within a sector. ADB appraises sample subprojects before the sector loan is approved. The borrower then selects and appraises additional subprojects during implementation. This makes the sector loan more cost-efficient than a project loan. It is also more flexible and adaptable to changes in situation, needs, and priorities.
Emergency assistance loan	An instrument for emergency support approved in the aftermath of disasters triggered by natural hazards, health emergencies, food insecurity, and technological and industrial accidents, and in post-conflict situations. It aims to support the efforts of DMCs to build back better and thereby enhance resilience under future emergencies. Focus includes early recovery activities, such as rehabilitating critical infrastructure and meeting basic needs, as identified by a post-disaster needs assessment, and reconstruction activities that may be needed to build back better. The emergency assistance loan may involve policy dialogue, particularly in post-conflict situations.
Multitranche financing facility	Supports complex projects that require a larger investment and longer commitment than a regular project loan could provide. ADB provides a series of tranches when the investments are ready and the borrower requests financing. The multitranche financing facility can finance multiple projects under an investment program in a sector or in various sectors, as well as large standalone projects with substantial and related individual components with long-term implementation plans.
Policy-Based Lending	
Policy-based loan	ADB will use PBLs to aid DMC structural reform, translating national climate policies into actionable steps. This flexible approach aligns short-term programs with broader goals, requiring sound policy frameworks for budget support. It includes TA to address policy constraints and strengthen institutions. A PBL provides tools for climate risk assessments and capacity building, supporting investment in climate mitigation and adaptation. It is the preferred funding method for developing and implementing climate policies in economic and social sectors. However, PBLs measure success and the processing of tranches primarily by output, such as policy documents developed, and are advised to shift toward increasingly measuring the mitigation and adaptation outcomes of climate policy action, which are mainly achievable in phase 2 of the PBL.
Contingent disaster finance	ADB will help eligible DMCs prepare fiscal frameworks against shocks and disasters through, for example, the CDF facility, a contingent financing option in some lending modalities, such as PBLs. Essential policy dialogue and reforms occur before emergencies, with disbursement when pre-agreed conditions are met. CDF provides immediate budget support post-disaster, focusing on legal and policy constraints for effective preparedness. Financing instruments like CDF and PBLs enhance climate resilience and environmental sustainability efforts. However, CDF deploys the concept of fiscal insurance against excess losses toward an annual cost premium, which may be hard for a tight DMC budget to pay.

Results-based loan	ADB can support DMC reform through RBLs, which tie disbursements to results via disbursement-linked indicators. An RBL enhances accountability and efficiency in government programs, promotes institutional development, and supports development coordination. It is particularly effective for linking fiscal frameworks to climate policy implementation and monitoring, as seen in the Paris Agreement's Enhanced Transparency Mechanism. RBLs also offer a framework for performance-oriented budget support and can increase the ambition and implementation of nationally determined contributions and long-term low-emission development strategies.
Combined Modality, Investment and Policy-Based Loan	
Sector development programs	ADB will deploy SDPs in DMCs needing both investment and PBL components, especially where policy reform incurs significant costs. SDPs integrate long-term sector needs, enhancing ADB's leverage for reform. They primarily involve public sector investment but may include private sector funds to stimulate responses to reform. SDPs are used where policy developments facilitate private investments, such as in renewable energy, transport, or infrastructure. However, SDPs by nature require increased up-front administrative efforts, notably to process investment safeguards.
Project Preparation Modalities	
Technical assistance	TA is exclusively by grant and given to help DMCs improve their capacity and make better use of development resources. Transaction TA supports a specific ADB-funded project. It may help to prepare an upcoming project, help deliver outputs or mitigate project risks under an ongoing project, or develop a specific public–private partnership project under transaction advisory services. Knowledge and support TA is not directly linked to an ADB-funded project. It focuses on knowledge sharing to build capacity, providing policy advice, and undertaking research and development. It can help develop a climate policy, a sector climate investment master plan, or a pipeline of future climate investment projects.
Project readiness financing	Project readiness financing comes in the form of an interest-free loan and finances project preparation and design activities for investments that are expected to be financed under one or more ensuing ADB-financed projects. Eligible expenditures include (i) consulting services for project preparation, such as detailed engineering design, feasibility study, and capacity building for future projects; and limited project startup activities, and (ii) pilot testing of project designs, including the cost of equipment and works.
Small expenditure financing facility	The SEFF helps ADB DMCs respond quickly and effectively to small financing needs. Once the facility has been established, the borrower can propose eligible activities for financing under it, and ADB can fund them as and when needed through a simplified process. Expenditures that are eligible for SEFF support are (i) consulting services to prepare or support the implementation of ADB-financed projects, such as detailed engineering design, feasibility studies, safeguard assessments, capacity building, and surveys for an impact assessment, and (ii) non-consulting services, works, and goods to support ADB-financed projects, such as pilot testing, rehabilitation, post-disaster early recovery, and operation and maintenance services. The SEFF has a fixed facility period of 5 years for a new commitment.
Private Investment Lending Modalities	
Financial intermediation loans	In partnership with the private sector, ADB will use financial intermediation loans to fund eligible financial institutions. These institutions provide green loans to sub-borrowers, such as small and medium-sized enterprises, female entrepreneurs, and low-income groups. Financial intermediation loans support decentralized renewable energy, micro energy solutions, and climate-smart practices for smallholder farms by enabling local banks to lend at their own credit risk.
Commercial private sector lending	ADB provides financing to private, state-owned, and sub-sovereign entities in DMCs. It offers hard and local currency loans, mezzanine financing, and equity investments, focusing on minority, noncontrolling investments. ADB also provides guarantees to help partners manage risks and uses blended finance to support private sector projects in emerging Asian economies, combining concessional capital with its own funds. Terms vary based on project needs and risks.

ADB = Asian Development Bank, CDF = contingent disaster finance, DMC = developing member country, PBL = policy-based loan, RBL = results-based loan, SDP = sector development program, SEFF = small expenditure financing facility, TA = technical assistance.

Source: ADB.

www.ingramcontent.com/pod-product-compliance
Lightning Source LLC
Chambersburg PA
CBHW041757220926
51847CB00014B/230
* 9 7 8 9 2 9 2 7 7 0 6 8 6 *